Children
Count

RETHINKING CHILDHOOD

Gaile S. Cannella
General Editor

Vol. 51

―――――――――

The Rethinking Childhood series is part of the Peter Lang Education list.
Every volume is peer reviewed and meets
the highest quality standards for content and production.

―――――――――

PETER LANG
New York • Bern • Frankfurt • Berlin
Brussels • Vienna • Oxford • Warsaw

MARY M. STORDY

Children

Exploring
What is Possible
in a Classroom
with Mathematics
and Children

Count

PETER LANG

New York • Bern • Frankfurt • Berlin
Brussels • Vienna • Oxford • Warsaw

Library of Congress Cataloging-in-Publication Data

Stordy, Mary M., author.
Children count: exploring what is possible in a classroom
with mathematics and children / Mary M. Stordy.
pages cm. — (Rethinking childhood; v. 51)
Includes bibliographical references and index.
1. Mathematics—Study and teaching (Elementary) I. Title.
QA135.6.S76   372.7'044—dc23   2014016403
ISBN 978-1-4331-1414-4 (hardcover)
ISBN 978-1-4331-1413-7 (paperback)
ISBN 978-1-4539-1385-7 (e-book)
ISSN 1086-7155

Bibliographic information published by **Die Deutsche Nationalbibliothek**.
Die Deutsche Nationalbibliothek lists this publication in the "Deutsche
Nationalbibliografie"; detailed bibliographic data are available
on the Internet at http://dnb.d-nb.de/.

The paper in this book meets the guidelines for permanence and durability
of the Committee on Production Guidelines for Book Longevity
of the Council of Library Resources.

*This book is dedicated to the memories of two amazing scholars,
Joe L. Kincheloe and Patricia Clifford, who left this
world much too soon.*

*Thank you, both, for being in my corner, and helping me to
see what is possible. I am forever grateful to have
called you my friends.*

TABLE OF CONTENTS

ACKNOWLEDGMENTS

It is with much gratitude that I wish to thank my former students of Grade 1C and Grade 2C from our school in the Rockies. These students are directly responsible for shaping my understanding of what children are capable of when given the chance. In particular, I wish to acknowledge 'Isabel' who was the catalyst in helping me learn to really listen to children's voices. I will be forever grateful for that conversation outside our grade-one classroom one October day.

Thank you to the hundreds of pre-service students I have had the pleasure of teaching over the past dozen years. Their questions, enthusiasm, and desire to make a difference in schools give me hope.

I wish to thank Jim Paul for his guidance and unwavering belief in me, David Jardine for teaching me about hermeneutics and introducing me to Gadamer, and Jo Towers who deepened my fascination with children and mathematics through her work and her actions. All three of these scholars have opened up my academic world. I wish to thank them for their patience, their time, and their thoughtful questions about this work.

Thank you to Laurie Bowers, an amazing friend, former colleague, incredible scholar, and avid sailor who has proved to be a valuable editor of my writing.

To my former colleagues and students at the school nestled in the Rockies, I am forever indebted. Directly influencing this book, I wish to acknowledge Sharon Friesen, who has been my mentor and friend since the first day I witnessed Pat Clifford and Sharon with children, and my former teaching partners and dear friends, Sherri Rinkel MacKay and Jeff Stockton, for their gifts of wisdom and laughter. I am deeply grateful.

I could never have undertaken the work that led to this book without the support and love of my family. I wish to acknowledge my parents, Marion and Gerald Stordy, for their unwavering belief in me and for fostering in me an intense curiosity about the world. Their sense of responsibility to others has impacted me greatly, and I can only hope I will one day be as generous and as giving as they are. My mother has proved to be solid as a proofreader and I am so grateful for the variety of ways she has supported my work.

To my Newfoundland friends, I am indebted for their encouragement. To my Alberta friends, Alex Fidyk and Bev Mathison, I am grateful for their unwavering belief in my work. To my Prince Edward Island childhood friends, Mary Norma Sherry and Bernard Lawless, I owe my sanity.

To the wonderful Shirley Steinberg I will be forever grateful. She is one of the most supportive and generous scholars I have ever encountered. I thank her for her mentorship and for writing the foreword to this book. I have much to learn from Shirley, and I look forward to many more years of learning from her how to live well in the challenging world of academia.

With love and much gratitude, I wish to thank my husband, Jody Doyle, for all that he has done in supporting my work, and for believing in me. I will be forever grateful for his compassion, his generosity, and his understanding. He is my true north. To my daughter Emma, thank you for providing me with an even greater urgency for change in our schools. I can only hope the possibilities discussed in this book may be given life in her future school experiences.

Finally, I must turn and acknowledge my beloved furry family—Sophie (1998–2014), Maggie, Austin (1999–2010), Katy (2002–2014), and Seamus—who have been steadfast beside me during the various stages that have led to the creation of this book. It is unbelievable what two cats and three golden retrievers have done for my soul. I can only hope, one day, to become the human they think I am already.

FOREWORD:
CHILDREN OFTEN DON'T COUNT ...
ESPECIALLY IN EDUCATION

Shirley R. Steinberg

I have told this story before, and sure I will tell it again ... because it needs retelling.

Sometime in the mid-90s, I was teaching my undergraduate course in Language and Literacy at Penn State. I was also kid-sitting for Henry Giroux and had his three boys hanging with me for the day. In the Chambers Education Building, there was a very cool hall, which stretched to another building, very sunny due to the windows, and mostly unused. At the end of the hall, by my classroom, was a pile of old furniture and, well, just cool stuff. The boys walked with me to class and spotted the furniture and the hallway. They started to feverishly build a fort; I left them there and told them I would meet them after class. As my students filed in, some of them stopped and spoke to the boys; I heard laughter and conversation outside the room.

About 30 minutes into the class, I heard a loud male voice, and the boys ran by the room and huddled in my doorway. Right behind them was Pete Rubba,[1] my department chair. He burst in the room, pointed to the boys, and queried: "This is a school of education, what are these kids doing here?" Pete twirled out of the dead silence in the class without waiting for our answer, and we were stunned. Seconds later, we collapsed into laughter.

Of course, our laughter was misplaced, indeed, Pete Rubba's comment was spot on ... there is no place for children in most schools of education. Mary's title says it all: children do count—both literally and societally—the catch is that most educators and, certainly, politicians don't count them. This book is Mary Stordy's insistence on children counting *in* schools and children counting *to* schools. She demands that we assure ourselves that we are doing our best to facilitate children to count in *both* ways.

Grounding her work on the notion that there needs to be a conscious shift from the archaic structure of schools into educational/mathematical settings which are meaningful, progressive, based on life-long learning and ways of knowing. Sounds obvious, but appears quite impossible in many contemporary public schools, and certainly in schools/faculties of education. Mary notes, citing Brent Davis, that mathematics should revolve around the human, not the inhuman, yet, again, the inhuman appears to be the only way most educational machines can function. Critical pedagogues Paulo Freire, Joe L. Kincheloe, and Henry Giroux have suggested for decades that *school is not for kids, because they don't count.* Mary's goal is to ensure that they do count.

Answering the primitive and positivistic cultural demands that mathematics students' work be recorded, re-recorded, counted with running records, and measured by standardized testing, Mary challenges teachers to create a human teaching atmosphere, one not directed by assessment and leveling. She insists that it is not by what we have to do, but what we should do—in mathematics, particularly—that we reconceptualize mathematics as part of our *lebenswelt*, culture, society, politics, and narrative. Setting an example of the accessibility of narrative, Mary's accounts of classrooms are compelling and complex, they also remind us just who it is we are teaching, their individual needs, passions, and ways of knowing.

Few books can recall classroom events, successful learning moments, with, ah, Aristotle, Gadamer, and Husserel, wound as theoretical threads within the weft of a discussion on authentic teaching. Mary Stordy does precisely this. With adept lenses of both teacher and scholar, this book exemplifies exactly the way mathematics education, and education in general *makes children count*.

PREFACE

Children Count is an interpretive exploration into the teaching of mathematics to children. Through the use of narratives to make meaning of pedagogic events, this is a book that considers the possibilities that exist for children and for teachers if mathematics is allowed to thrive in schools as a living human enterprise. Such a reconceptualized view of mathematics challenges the status quo and results in a different image of schooling. It is an image that is messier, requires more of the teacher, and expects more from students. President Barack Obama wants the American educational community to "educate to innovate." The struggle many teachers and parents have is that they are not sure what that phrase actually means in action. Science, Technology, Engineering, and Mathematics—STEM—has become a familiar term in the United States and Canada in the past few years, and is quickly catching hold in other parts of the global community. The following chapters you are about to read had elements of STEM teaching and learning with younger children long before the term was really known. What needs to shift in schools framed in an industrial age—despite all that we know about learning and children—that now exist in a time in which inquiring into STEM education has political and economic thrust and the power to transform the kinds of thinkers our schools can foster? While the intention of the text is concerned primarily with mathematics teaching and learning, elements of STEM teaching and learning can be seen woven throughout the narratives.

I wrote this book because I have realized from my work with pre-service teachers there is a need for texts in education to provide images of teaching that are hopeful as well as academically rigorous. Such images of mathematics teaching are somewhat rare. When it comes to educational research texts, pre-service and practicing teachers read about methods, about inquiry-based learning, about empowerment of children and teachers, about academic rigor, about holism, about integrated curriculum, and about generative curriculum, but they are often not sure what all this might actually look and feel like in a real classroom with real children. *Children Count* attempts to give the reader images of what such a classroom might look like. It also allows the reader the opportunity to gain insight into my professional and personal growth as an educator who transformed from a teacher who did not wish to stray too far from traditional practice to one who strives to embrace teaching and mathematics as living disciplines. The text captures my mistakes, my choice of actions, and my decision-making process from the unique perspective of a teacher who reflects and learns from my students when I realize I *must* listen to them, because what they have to say counts.

In an age in which reactions to international test score rankings in mathematics become front-page news and media fodder for weeks, this text might provide a refreshing change to another way of thinking about being with children and with mathematics in schools.

· 1 ·

BEGINNINGS

Out in the Hallway

"I didn't know it was going to be so hard," the young child said, standing outside of the classroom. Her arms were crossed, her brow furrowed, and her bottom lip jutted out. Her dark eyes met mine. Then she looked away.

"What?" I asked, not sure where we were headed.

"School. If I had known it was going to be like this, I never would have come out." I bent down and squatted next to her, carefully adjusting my skirt in the process. I placed my hand gently on her back and tried to meet her eyes.

"Come out? You mean come to school?"

"*No! I never would have come out of my mom if I knew it was going to be so hard!*"

Before me stood a six-year-old girl whom I was just getting to know. For almost two months we had been in each other's company, and she seemed like she was a bright girl so the schoolwork couldn't have been 'hard,' as she put it. What could she mean?

The date was October 21, 1998. It was early in the fall of grade one, and in many respects the girl, Isabel, was trying to come to understand how to do school and there were things about it she appeared not to like. The part about doing things when the teacher [me, Mary] asked her was one of those things Isabel seemed unsure about. My hallway conversation with her outside our noisy classroom turned out to be one of the most important conversations that I would have as a teacher.

Upon moving to western Canada in 1998, I accepted a teaching job near the end of August of that same year. School started only days later. Up to the time of the hallway conversation I had been going flat out, learning the provincial curriculum, teaching grade one for the first time, and living in the rugged foothills of the Canadian Rockies; all these topographies lay in contrast to the pristine rolling fields of the Atlantic island where I had begun my teaching career.

Every day in my new home was a day of survival. I was a complete foreigner trying to understand the lay of the land. Cougars, bears, and wolves were residents of the forests surrounding the school. Children were not to eat food outside for fear of attracting animals to the playground. I had never before encountered this issue. As a classroom teacher I had to accommodate both a snack time and a recess time. It all felt so new, but I was not a beginning teacher. I had been teaching in two other provinces prior to this job, but teaching here felt new. It felt as if I were freshly plucked from a teacher education program and I was experiencing the ultimate test of survival.

The school was a public community school that had been built a year and a half before my arrival. Teachers had been brought together to form a staff that was encouraged to challenge the status quo and create new images of teaching practice. I was hired because of my graduate work that had a focus on teaching and technology. I was one of two grade-one teachers and we were both new to the school. My teaching partner was entering her eighth year of teaching and the administration made it clear that grade teams were expected to work together. The principal specifically asked me to help my grade partner integrate technology into the classroom. In turn, she was to help me adjust to teaching the province's curriculum for grade one. For the first few months I had been careful not to step too far out of the box and away from my fellow grade-one teaching partner.

Standing out in the hallway with Isabel, I suddenly did not recognize myself. Even what I was wearing seemed foreign. I do not wear skirts! When we stepped back into the immediacy of the goings-on of the classroom on that October day, I saw my classroom differently. I remember standing completely still and looking around, unaware of the buzz of the children. I gazed at the walls of the room. Where did all those laminated apples come from? How can red be that shiny? My attention turned to the commercially produced calendar corner. What was that about? And who said that I had to do 'calendar' every single morning?

The conversation with Isabel out in the hallway shook me to my core. "If I had known it would be this hard, I never would have come out." Her words were out there. They were loose in the world. They jarred me. In two months had I managed to make school so terrible that she wished she had never been born? I did not know I had settled into an easy doing of things, of disappearing into the doing of things in a school with children.

So now what was I to do?

This child had spoken to me. And I needed to listen, to really listen and pay attention to what she had to say. What was it that we were doing that felt so hard? Clearly Isabel did not feel she belonged in the place of school.

As I watched this young girl walk slowly back to her desk to finish her work, I realized that I didn't recognize anything anymore. With our talk out in the hallway, Isabel had transferred her suspicion about schooling to me. I felt like a stranger in my own body. Had forgetfulness guided me away from the center of my own being? But what had I forgotten? My intentions were to be a good grade-one teacher. I only wanted to do things right. I wanted to fit in with my grade team. I wanted to belong. I lingered on the question, What had I forgotten?

The very night of my talk with Isabel, I began to pore over the *Alberta Program of Studies for K-9 Mathematics* (1996) for grade one. There were three specific outcomes that related to using a calendar such as: "Sequence events within one day and over several days; Compare the duration of activities; Name, in order, the days of the week and the seasons of the year" (p. 30). My students were already doing that with ease. Besides, were there not many ways of addressing these outcomes other than through a morning routine that I directed? Our calendar routine took up more than three hours of instructional time each week.

I needed to find a path that would invite me to dwell with these children, and not merely navigate a classroom toward the already determined destination of grade two. So, could I be the problem? Had I lost my imagination? Might that be it? Was I so paralyzed by doing the right thing at the right time as a teacher that I forgot to be all right with children?

When I first came across Gadamer's writings several years ago I felt his words speak directly to me and my work with children. Gadamer (1989) suggests that everyday life is constituted through forgetfulness. Human life is the process of leveling out and flattening everything. Looking back, I see that I had flattened the landscape of that grade-one classroom into a reduced pile of laminated apples and months of the year. My way of doing school had to change.

In *Truth and Method* (1989), Gadamer claims that the happening of events is essential for understanding. In life things happen. What Isabel had said to me happened. Her words called to me and those 'words' would not go away. How was I to proceed as her teacher?

Gadamer (1989) writes about the importance of listening to belonging, noting:

> In human relations, the important thing is, as we have seen, to experience the Thou truly as a Thou—i.e., not to overlook his claim but to let him really say something to us. Here is where openness belongs. But ultimately, this openness does not exist only for the other person who speaks; rather, anyone who listens is fundamentally open. Without such openness to one another there is no genuine human bond. Belonging together always also means being able to listen to one another. (p. 360)

Whether I wanted to or not, I could not, according to Gadamer "hear away" (p. 462) what Isabel had said. I needed not to 'look away' from the place I had formed for my grade-one students.

I needed to listen. These children had something to say to me about how grade one could be. In retrospect, I was lucky to be teaching in a place that allowed me to work with a few truly experienced teachers, who were open to children's questions and who listened to children's voices. I felt myself gravitate toward these staff members almost immediately after my talk with Isabel in the hallway. And I began to sit with my question: What had I forgotten? Could it be that I had forgotten about attending to children even though on the surface that is what it appeared I was doing as a classroom teacher? Could I have forgotten to attend to the topic at hand? Did I forget about myself as a learner? Had I forgotten about the wonderful possibilities that can exist in a space when wonder and curiosity are invited to flourish? I needed to pay attention to all of these things.

The Post Office

Before the same school year had begun I had decided that I wanted to make reading and writing relevant to the children in my grade-one class. It was now early November. On this day we sat on the floor in a circle. All twenty-three of the children were in attendance. I waited for the room to get quiet.

"Well, grade one, you have been working really hard at writing in your journals ever since our first day of school."

"I'm up to a whole page now, Ms. S. And that's without the picture!"

"Wow! Good for you Justine. Can anyone tell me why it is important to be able to write?"

"I know! So people know what you are thinking."

"That's an interesting way of putting it, Malcolm. Anyone else?"

"Well, if you can't talk to someone, like my dad when he is working late, you can write a note. I write my dad a note so that I can say 'good night' to him even though he is not there before I go to bed but he reads it when he gets home." [Inner Voice: Cool. Almost have them at 'writing letters' already. We are ahead of schedule. I wonder which book we should read before recess?]

"Does anyone have an example of when you see your mom or dad writing?"

[Inner Voice: Look at all those hands shooting up around the circle. They love this!]

"William?"

"My grandmother writes letters! We get them at the post office."

"Yeah, Ms. S, can we write letters?"

[Inner Voice: All right! My plan is working! I'll have them convinced in no time to decorate milk cartons that they will bring from home, and then we will staple them together into a compact square consisting of six rows of five.]

"Hey, can we? Maybe we could write letters to each other!"

"That is a great idea, Hannah. You too, William. Maybe we could use milk cartons as our own mail slots. Did you notice in Ms. T's classroom how they have those little boxes right by the door? We could mail each other letters using our own mailboxes."

[Inner Voice: This is going smoother than I thought! They are all quiet and attending to my every word right now. I wonder where we should put the mailboxes. Maybe over there, by the pencil sharpener?]

"Why can't we make our own post office?"

[Inner Voice: Whoa! Where did that come from?]

"Can we, Ms. S? Can we make our own post office?"

[Inner Voice: Make our own post office? We don't have that kind of time! I didn't plan for this! Who builds a post office? Maybe I should just pretend I didn't hear what Joey just said. That could work. Is it time for snack yet?

We could just read a book. I bet they'll forget all about this by the time recess rolls around.]

"Can we build a post office?"

[Inner Voice: Hold on, Mary. You have a choice here. You can talk them out of this. You've done it before. But wait. Just wait. Think about it. Maybe there is something to this. Look at them! Look at how excited they seem. Even Marcia is anxiously leaning forward, waiting for an answer. You've wanted to do things differently, remember. And Isabel's staring right at you. Say something!]

"You want to make a post office? Well, what are you thinking about exactly?"

"Maybe we could make it out of cardboard or something and it could have a door and all our mailboxes could be inside of it, like a real post office."

"Yeah! Ms. S, we could paint it and put it over there in the corner."

"And we could even write letters to the principal too and give him his own mailbox and he could write to us!"

[Inner Voice: Oh boy, I hope I don't regret this.]

"Well, okay. If we are going to do this, we will need a plan. We have lots of work to do!"

"Yay! This is going to be so cool! Thanks Ms. S!"

I didn't regret it. During the next week, we built a post office. It was made of cardboard with a black roof that we suspended from the ceiling using paper clips and string. The walls were painted red and black in brick-like fashion. It had a door that actually opened and closed. Inside the post office were all the mailboxes, individually designed out of milk cartons. There was even a box for the school administrators. We had many conversations about when the post office should be open and when it should be closed. We decided that we would need to make a sign that signified whether it was open or not. In true six-year-old fashion, the sign was printed in multiple-sized letters and had two sides to it that we flipped around depending on the time of day. When the post office was open I would often hear stifled giggles emanating from inside of the 'building' as children checked their mail. They seemed to love their simple creation. Letters were written during recess, at home, during free time, before school, and after school. I had not anticipated how many responses I would have to make. I guess I had not thought they would write as much and as often as they did. We created rules of etiquette when writing our letters, such as what length of time would be appropriate before expecting a response. Even the principal and assistant principal became a part of the action and made regular trips to our post office to mail letters to the children. When school information was sent home, they were first distributed through the post office. The child-made creation in the corner of the classroom did so much more than prompt the children to write. It signified something that was truly theirs.

As I think back to the moment when the student Joey first brought up the idea of building a post office I remember the question more as a statement. It was as if he did not expect to be heard. Did he not expect an answer? What if I had dismissed his idea as I had thought of doing? When I think about how much stronger my relationship became with the children during our work together on the post office, it scares me to think I came so close to not seeing the possibility when it presented itself. The planning and construction as well as the physical act of going to, going in, and doing the mailing at a post office were as much of the experience as the writing of the letters.

Clifford and Friesen (2003) write of imaginative engagement as "the kind of engagement that invites children most fully, most generously, into the club of knowers; not at some unspecified time in the future when they are grown up and able to use their knowledge, but today and each and every day they spend time with us" (p. 22). In letting go of my own plan and accepting the idea of the post office, I had allowed the children into the "club of knowers" (Clifford & Friesen, 2003) for my first time. They designed it and had a say in how it was to be used. That homely post office was an act of imagination on the part of the students, and we all delighted in using it. It was theirs. It was ours.

Keeping Up with the Children: Journal Reflections

Thursday, January 14, 1999

The children really are growing! Justine can barely get into her desk. Today I took all the books that the grade seven teachers, Bonnie and Karen, gave me last week, and I spread them out on the floor of the classroom for the children. I am really lucky to get to use these children's books of legends and myths from the medieval time period. Bonnie and Karen have been teaching for a long time, so I guess it only makes sense that they would collect books over the years. Even Jace couldn't get enough, so that was awesome to see. I am not surprised that the book on armor and weapons already fascinates him. I am not exactly sure where this work is headed, but when David kept saying how cool it would be to live back then I knew that we needed to spend some time figuring out whether he is right. In fact, almost all of the children thought it would be neat to live in a time of kings, knights, queens, and castles. So we are at the beginning of finding out more about what life might have been like back then. I need to spend some time with the curriculum to see what I can integrate.

Friday, January 22, 1999

This afternoon Kelly suggested that we should make castles. That is a great idea, and it will be just the way to integrate some mathematics into our study of the middle ages. We could use sugar cubes because I remember seeing that at my last school back east. Every child could make his or her own or maybe work with a partner? I'll have to figure that out. But I know that I need to address spatial sense, so this is one way to enter that. And there would be some estimation, problem solving, and comparing. I'll have to remember to get the parent helper to take pictures for student portfolios. That could work! Hey, this teaching thing is not so difficult after all. I'm integrating curriculum and giving the children hands-on experiences. I can't wait to tell them tomorrow about our sugar cube castles. I bet they'll love it!

Tuesday, January 26

Well, the whole sugar cube castle didn't go over the way I had hoped. In fact, they weren't really that interested. They got it into their heads that we should make one big castle that we can all enter. They said castles have drawbridges and that we should build a round table too, like the one that King Arthur had. I need to think about this some more.

Wednesday, March 10

The two grade-seven boys came down to our classroom today and cut the donated plywood. The kids and I had figured out the size that we would need to fit everyone around the table considering we had two sheets of wood that were each 4 feet × 8 feet. Of course we had to really think about how to make the table portable, and together we came up with the idea of using hinges so that the table could fold into a half circle. I think Bobby suggested it. I have no idea how the two grade-seven students managed to take a circular power saw to school in their backpacks but they did. The noise from the saw on the wood was pretty loud but the other classes tolerated it. Now the children want to paint the table exactly the same as the way it is in one of our books. I wonder how it will turn out.

Thursday, May 20

Today a chef who specializes in cooking food from the medieval period brought us five dishes to try. We learned so much about why fruit would be mixed with the meat and about dental hygiene back then. The chef loved the day almost as much as the children!

Sunday, June 20

Well, tomorrow is our big event. I managed to move the entire castle down to the school gymnasium. I'm exhausted. I hope the children remember their lines! The purpose of our 'Medieval News' production is for the children to have a chance to share with the school and their families what they have figured out during our work on the Middle Ages. Writing the script with them was so comical. They loved the movie Monty Python and the Holy Grail (the edited version). It took me ages to split those coconuts for Tyler. I can't believe they want to gallop using the coconuts for horses just like Monty Python. I know that Kelly and Joseph will be great news anchors and will ad-lib if needed in between the reporters' stories. Our feast after the performance on the eve of the summer solstice is only fitting. I am so surprised by how hard the children have been working on the castle, on the round table, and on their own areas of research. I never would have thought it could be like this. I am so proud of them. And we moved far beyond the Program of Studies *outcomes suggested for grade one. I cannot imagine if I had said, "No, we can't figure out the pulley system for the drawbridge because you are supposed to learn pulleys in grade two. Sorry." Mr. Allan officially told me that I will be able to continue on with this group of children into grade two. I can't wait.*

Feels Like Home

So we sat under the Martian sky, with Earth a mere blue sphere in the distance. All twenty-eight[1] of the children gathered on the floor with me around a red, scaled-down section of the terrain of Mars that we had created with the help of a university mathematician. Lego robots peppered our Martian surface. A tiny portable video camera sat slightly askew on one of the robots at the base of Olympus Mons.[2] It was hard to believe that June 2000 had finally arrived. Sunlight shone through the window and onto the castle remnants created last year when the children were in grade one. On the other side of the classroom a breeze entered the room through the back door held open by an old shoe. The breeze carried with it the distant sounds of children's laughter from the playground. The walls of our grade-two classroom were covered with student work. From where I sat I could see the Greek gods and goddesses, moon study art, class conversations transcribed in print, instructions for caring for our class snake, and interesting ideas that were figured out about programming robots. Behind me our snake, Elaphe[3], was poking his head out between the totem pole and the flat rock he would hide beneath. The SMARTBoard[4] had been nudged to the side of our small pod of computers and was now shoved against some desks. All twenty-eight individual desks were pushed out of the way at the perimeter of the classroom. They had simply gotten in the way of our work. I sat in a pair of denim overalls with my legs crossed. The blue round table poked out from underneath the Martian surface. The children were quiet. The vibrations from the buzz of the heat lamp on the terrarium were rhythmically soothing. I felt such peace just being present with these children. They had taken me on such an adventure for two years. I remember smiling and catching Isabel's dark eyes to the right of me. She smiled right back, and spoke:

"It feels like home here, Ms. S."

Only two Octobers ago she resented being born. I let her comment drift over me. This child. Her words.

Heidegger (1977) writes that to dwell means to be "the way in which you are and I am, the manner in which we humans are on earth.... Dwelling itself is always a staying with things" (p. 329). He suggests that "even when we relate ourselves to those things that are not in our immediate reach, we are staying with the things themselves" (p. 334). Perhaps Isabel's words were about the manner in which we dwelled with one another. In looking back on the experience noted above, I am struck once again with the power of her words; in sharp contrast to her statement in grade one, these words were about feeling at home. Since creating the post office, I had tried to be more present to the children and to dwell with them. I slowly realized that children have something to say about how to be. Winning (2002) builds on Heidegger's "staying with things" and writes about the space of home:

> The things from home and the space back home are connected and so can give us a sense of home. Therefore, we use the idiomatic phrase: There is a little bit of home in this. We bring things from home to put up around us in the new abode because in the things themselves there is the space of home. (¶ 10)

I too felt the space of home with Isabel's class. I felt it in my core.

· 2 ·

BEING LED ON

Possibilities

"Mary! Can you believe we are starting another school year already? How was your summer?"

"Great! I am so excited to be back, though. Are the buses here yet?"

"No, not yet. Hope you will enjoy teaching grade five!"

"Thanks! I'm looking forward to it!"

It was the start of a new school year. New grade. New students. New possibilities. I stood back and critiqued my handwriting on the whiteboard. As a left-hander, today's date, 'August 31, 2000,' contained a definite slant in the bright blue marker ink. The start of the new year meant all markers were fresh and full of color. The sharp scent of it filled my nostrils and made me slightly light-headed. Perhaps it was the anticipation, the excitement I was feeling that led to my light-headedness. Soon, thirty children in grade five would be coming through my doors filled with energy and stories of their summer experiences. I thought about all I had learned since arriving at the school two years before as a new grade-one teacher. I shot a quick glance around the room. Not a laminated apple in sight. Having lived and worked with those wonderful children from grades one and two, I knew that I wanted to help create a class of students filled with wonder and curiosity as I had done last year. I knew it was important to invite the students into the work they would do. "This is going to be such a great year," I whispered to the whiteboard. "The students will be even more mature and articulate by grade five. We will be able to go really far with our investigations. There are so many rich possibilities connected to the grade-five curriculum!" I thought back to conversations I had with parents last year who urged me to consider moving up to grade five to work with their children. I opened the window and heard the rumblings of buses. Almost immediately the hallway exploded with shouts, laughter, and a single tut-tutting from an adult. I smiled. They were here! I smoothed my cords and tugged at the sleeves of my blue cotton blouse. They began to pour in, and quickly the room filled up. 'They're so big!' I thought. At the same time, I couldn't help but wonder about my grade-two students from last year, entering new classes up the hallway from my own. My new students, the ten-year-olds, came in and sat down in new fall outfits, too warm for today's August sun. I waited for the rustling to settle. I thought, 'This is going to be so great! I know it!' I quickly introduced myself and rearranged the students so that everyone could see one another for the upcoming conversation. When everyone seemed ready, I spoke:

"Who are you? What do you think about? What interests you? What is important to you?"

I looked at each face as I spoke. Then I waited. The room remained completely silent.

Being Led On

Following my one year of teaching grade five, I was recruited as a consultant for the school division. From 2001 to 2003, I worked in six specific schools and helped teachers who were wrestling with inquiry-based teaching and learning as well as technology integration. In this role, I would team-teach; other times I would meet one-on-one, in small groups, or hold large-group meetings with teachers. The work was challenging because inquiry-based teaching and learning are not easy. My goal was to be provocative and respectful. I tried to ask thoughtful questions of teachers about the reasons they did what they did with children. I tried to point to classroom events that would be worth taking up. I spent a lot of time in grade-one classrooms and eventually I began to feel overwhelmed. In some rooms I noticed entire classes of children who were silent. I watched them at recess time and they would be playful, but entering the threshold of the classroom they would put on their indoor shoes and get ready for whatever the teacher had for them to do. I found this unsettling, because there had to be children like Isabel in those classrooms.

My work as a consultant gave me the opportunity to work closely with some university researchers. After many conversations with them about teaching, inquiry, and children, I decided to pursue a doctorate to conceptualize my work with children.

At the same time that I was beginning my doctoral work I also began a two-year secondment at the same university. This meant I was teaching fulltime in a teacher education after-degree program. While the work kept me very busy, it also allowed me to become exposed to many rich experiences. In 2003–2004, I spent time in nine schools and thirty-one classrooms across the city supervising student teachers. Teaching in the education faculty also gave me the chance to be surrounded by students who were beginning the process of figuring out who they wanted to be as teachers. In fact, my work with student teachers actually led me to my topic for further study.

Blindsided

From my university office window on the twelfth floor I could see people scurrying in from parking lots to escape the January cold. Maybe some of those moving dots would be students of mine. Today I would meet a new group of education students who would be taking up a new curriculum course. In preparing for teaching this course, I had imagined how the semester might work. I was excited to be taking up the many facets of curriculum inquiry. I had cases from my own teaching I wanted to use. I had artifacts and exemplars of children's work. I had a number of readings and even some mathematics manipulatives. I was ready for anything.

Ten minutes later I loaded up my bag and headed to my assigned classroom. In the elevator I looked over my class list with seventeen students' names. There were two men and fifteen women in this curriculum inquiry into elementary mathematics, science, and physical education. When I got to the classroom the students had already gathered in a horseshoe formation. The only place for me was at the empty space at the head of the horseshoe. As at the beginning of every class I teach, I introduced myself briefly and then asked the group almost the same questions I had asked my grade-five class. The conversation ran as follows:

MARY.   Now I want to know, when it comes to mathematics, what is it that interests you? What do you think about? What do you feel is important?

	I know we are just beginning this work, but you have been in classrooms for one semester and you also spent your childhood taking up mathematics in school so you probably have lots of experience to draw from already.
	At first the room was quiet. Then a dark-haired woman, Jessica, spoke up.
JESSICA.	I have a minor in mathematics and I have always loved math. I got good marks growing up and I took math all through university. Right now I am doing my field placement at an arts-based school, so one thing I have been thinking about is the connection between art and mathematics. I really want to learn more about that.
MARY.	That is great, Jessica. The connection between mathematics and art is fascinating. [Inner voice: Wow! This is turning out just like I hoped it would!]
	Then Angie spoke.
ANGIE.	I hate math. Teachers made me feel dumb because I took so long to do problems. By the time I got to high school I was barely passing. My mom had to get me a tutor. I never took another math class and I have no idea how I am going to teach math to kids because I don't understand it myself. So I am kinda freaking out. I really don't want to do to students what was done to me.
	I nodded a slow nod that meant 'I hear you' to Angie. [Inner Voice: Uh oh. What is going on here?] Before I could respond, another student continued.
STUDENT 1.	Yeah, me too. I remember my stomach hurting when the teacher timed us for Mad Minutes[5] in grade three. I used to tell my mom I was sick, so I wouldn't have to go to school. I sucked in high school math. I had to use my French mark to get into university, 'cause I never did math in grade twelve.
	Again, before I could get a word in, another student added something.
STUDENT 2.	They do those Mad Minutes in my classroom for field. Some of the kids just sit there. I wanna help them but I'm not supposed to. It's kinda sad, really.

Like weavers bringing in new threads, students linked story after story of how they were scarred by a mathematics teacher who made them feel stupid, and by a discipline that seemed completely foreign to them. They expressed their anxiety over having to make mathematics interesting for children. Some brought up recent events in their field placements that sounded more like classes during the seventies 'back to the basics' movement for teaching mathematics. The student next to me, the last to speak, announced: "I'm gonna have to fake it, and kids can sense when you're faking it." She turned to me and added in a lower tone, "Dr. Jarvis talked about that in lecture."

Then to the rest of the class she said, "I'm afraid that they'll end up hating math too."

I was not prepared for this. I was completely thrown by the students' words. Was it just a freak coincidence that out of seventeen first-year post-degree students, eleven were terrified of mathematics, five felt they were high achievers in the subject but didn't really like it, and only one felt confident? Surely this was unusual.

The following morning I had another group of students for the same course; these students I knew from the fall semester. I asked them to talk about their experiences with mathematics and what they hoped to gain from our time together. The fear, lack of confidence, anxiety, and dislike were almost identical to the first group of students' testimonials

I had been blindsided. Why did I not see this coming? Until this point, I had lived in schools for ten years as a teacher. How could I have experienced my own education and teaching experiences in other parts of Canada and not realized the culture surrounding the discipline? Had I known and forgotten? Why had it never been revealed to me this way before now?

Were my students just poor in mathematics? How could it be that elementary generalist student teachers saw mathematics the way that they did? Did they not realize how culturally significant the learning of mathematics is for children? Did they think they could somehow escape teaching the subject? Why did they want to become teachers in the first place if that is how they felt about such a key discipline in schools? How might I enter this chasm between what I perceived to be at hand and what really was at hand? What was I going to do?

Opening Up

Understanding is an adventure and, like any other adventure it is dangerous, ... but when one realizes that understanding is an adventure, this affords that it provides unique opportunities as well. It is capable of contributing in a special way to the broadening of our human experiences, our self-knowledge, and our horizon, for everything understanding mediates is mediated along with ourselves.
    H. G. Gadamer, *Reason in the Age of Science* (1981, pp. 109–110)

Coming to understand something hermeneutically is less about getting to an answer or the solution and more about exposing what might be possible and what may open up to the reader and to the writer of a hermeneutic inquiry. Davis (1996) suggests, "Indeed, the hermeneutic question might better be thought of as an issue or topic of wonder" (p. 25). Questions present themselves to us when we are open and ready to see them. Much like Isabel's statement, "If I had known it was going to be like this, I never would have come out," presented itself to me, so too did the response of my student teachers toward mathematics cause a rupture in my understanding. The happening of events is essential for understanding. What my pre-service students had said, happened. Gadamer (1989) explains:

> We have already seen that, logically considered, the negativity of experience implies a question. In fact we have experiences when we are shocked by things that do not accord with our expectations. Thus questioning too is more a passion than an action. A question presses itself on us; we can no longer avoid it and persist in our accustomed opinion. (p. 366)

I had found my topic of wonder. I could not ignore it. I began to read, to talk, and to dwell with the complexities that the students presented in that first math education class. The questions and the strong feelings of my elementary generalist student teachers in that first class in January 2004 has led me on to my topic for further inquiry. In the following chapters I wish to unpack and live in the spaces of the complexity of teaching mathematics to young children. My desire to understand the teaching of mathematics is twofold: I want to come to understand better the complexities of teaching mathematics to children and in doing so I want to lay bare my own awareness of the difficulty that seems to have been forgotten and remembered, revealed and concealed. Had I completely overlooked the complexities of teaching mathematics? In the foreword to the second edition of *Truth and Method*, Gadamer (1989) writes, "My real concern was and is philosophic: not what we do or what we ought to do, but what happens to us over and above our wanting and doing" (p. xxviii). The writing of this book will help me understand what has happened and is happening to me as a teacher of mathematics.

This text will consider the relationship of mathematics to teaching in terms of the past and the present, the particular and the general, the philosophical and the practical, the part and the whole. It is an exploration into what might be possible when it comes to teaching mathematics to children when the world, which includes the living world of mathematics, is allowed entry. Jardine (1994) describes mathematics as not being held in a fixed state, and he goes on to write:

> It is, so to speak, a way which must be taken up to be a living whole. There is thus a way to mathematics. Learning its ways means entering into these ways, making

these ways give up their secrets—making these ways telling again, making them more generous and open and connected to the lives we are living out. (p. 270)

What may be possible for the teacher of mathematics if it is thought of in this way? Possible for the child? Possible for mathematics? What might it mean for pre-service teachers to be ready to teach mathematics to children? What does it mean for any of us to be ready to take up mathematics with children? Embracing Gadamer's (1989) idea of "the fecundity of the individual case" (p. 38), this text will explore my own lived experiences with children, teachers, and pre-service teachers to come to a deeper understanding of the teaching of mathematics to children

Having taken Gadamer's (1989) words to heart about experience, about tradition, and about our existence in the world, I realize that I need to take up the same questions that I asked my own grade five students and apply them to myself: Who am I? What do I think about? What interests me? What do I think is important? What would I like to do? The anecdotes that are present in this work show in some ways who I am and who I am becoming, as well as some of what I think about and hold important.

In the introduction to *Truth and Method*, Gadamer (1989) writes, "The way we experience one another, the way we experience historical traditions, the way we experience the natural givenness of our existence and of our world, constitute a truly hermeneutic universe, in which we are not imprisoned, as if behind insurmountable barriers, but to which we are opened" (p. xxiv). As the quote from Gadamer at the beginning of this section states, I realize that this inquiry explores the process of undergoing an adventure that is dangerous but that it also provides unique opportunities that such understanding brings forth. I welcome you to this adventure.

· 3 ·

RECONCEPTUALIZED MATHEMATICS

Mathematics teaching is … not amoral, as it claims, but indisputably immoral. In allowing itself to forget that its subject matter is a humanity, it has become an inhumanity. It is thus that we have created a system that values compliance over creativity, that spawns destructive behavior by destroying our experience, and that conditions learners to reach for the formulae ahead of the imaginative. (Davis, 1996, p. 281)

The language in the above quotation by Brent Davis, past Canada Research Chair in Mathematics Education, seems, at first, strong. The subject matter of mathematics is actually a humanity, and now it has become an inhumanity. How has it come to pass? This chapter takes up the current nature of mathematics, that is, the way mathematics has come to be viewed in the world by both insiders and outsiders. I will begin by exploring the more commonly held public view of the nature of mathematics and the difficulties that it poses for mathematics educators and classroom teachers. I will then draw on the work of the reconceptualists in mathematics education to offer a different view—a more pedagogic view of the character of mathematics.

A Public View of Mathematics

There are many entry points for exploring the public view of mathematics. The traditional notion of mathematics as a body of knowledge that is objective, absolute, certain, and incorrigible rests on the foundations of deductive logic (Ernest, 1996; see also Applebaum, 2003; Davis, 1996; Friesen, 2000; Lakoff & Núñez, 2000; Paulos, 1991; Sam, 2002; Taylor and Sinckir, 2000) and leads to the widespread public image of mathematics as "difficult, cold, abstract, theoretical, ultra-rational, but important and largely masculine" (Ernest, 1996, ¶ 5). Lakoff and Núñez (2000) refer to the absolutist view as the mythology that surrounds mathematics. They argue that the idea of math as an objective feature of the universe is attractive; mathematics as "queen of the sciences" defining absolute precision is sexy (p. 340). Lakoff and Núñez refer to this vision of mathematics as the "romance of mathematics." In this romantic view, human interaction does not change the nature of mathematics—this disembodied and abstract phenomenon—since it is objective and referenced on an external universe (Davis, 1995, 1996; Ernest, 1996; Friesen, 2000; Jardine, 1998; Lakoff & Núñez, 2000). Lakoff and Núñez critique the mystique of the one who works with mathematics, the mathematician, based on the premise of the myth: "It [the romance of mathematics] perpetuates the mystique of the Mathematician, with a capital 'M,' as someone who is more than a mere mortal—more intelligent, more rational, more probing, deeper, visionary" (2000, p. 340). Pythagoras and Descartes are two mathematicians whose contributions to the shaping of the traditional view of mathematics have been fundamental.

Mathematics as a Secret Cult

Pythagoras was the leader of a secret cult of Greek mathematicians. There is historical speculation concerning the reason the mathematicians worked in secret. Perhaps it was to keep the knowledge from others. Knowledge is power. Other historical accounts of the Pythagorean School's activities were based on the secrecy stemming from Pythagoras's experience with the secret practices of Egyptian priests. Revealing revolutionary ideas could cause trouble and stir up opposition (Mlodinow, 2001). "One of Pythagoras' discoveries became such a secret that according to legend, the Pythagoreans forbade its revelation on penalty of death" (Mlodinow, 2001, p. 25). Mathematics was deeply embedded in the spiritual and mystical lives of the ancient Greeks. The elegance of mathematics and the understanding of mathematics "hinted at an ultimate 'perfection'" (Davis, 1996, p. 66) and, in turn, a being closer to God. During the time of the ancient Greeks, mathematics was considered "a particular mode of inquiry" (Davis, 1996, p. 66), embedded in explaining the universe and the physical nature of objects on and around Earth.

Descartes's Influence

It was during the time of Descartes and his contemporaries that a major shift in the discipline took place. Breaking away from tradition and the connection of the mind and body, "I think, therefore I am," Descartes and his model of reasoning gave birth to empiricism and rationalism as the "key to knowledge" (Davis, 1996, p. 4). Mathematics became separated from the world to become an isolated discipline. Mathematics also became separated from human experience and gained the status of being the avenue of reasoning. Mathematics became the path to scientific truth. It was elite.

Lakoff and Núñez (2000) argue that today, thanks in part to the seventeenth-century shift in the field, mathematics intimidates people and the romance "serves the purposes of the mathematical community" by creating, maintaining, and justifying an elite (p. 341). The authors go further and claim the following:

> It [the romance of mathematics] is part of a culture that rewards incomprehensibility, in which it is the norm to write only for an audience of the initiated—to write in symbols rather than clear exposition and in maximally accessible language. The inaccessibility of most mathematical writing tends to perpetuate the romance and, with it, its ill effects: the alienation of other educated people from mathematics, and the inaccessibility of mathematics to people who are interested in it and could benefit from it. (p. 341)

School Mathematics

School mathematics has perpetuated the myth of a mathematics that is absolute, objective, fixed, and static. Empirical studies have shown that teachers' beliefs about mathematics influence how they teach the subject (Ernest, 1996). Friesen (2008) claims there is no other school subject in which the body of knowledge that makes up the discipline in the world is so far removed from the curriculum that students see daily. According to Paulos (1991), mathematics in schools gets identified with "a rote recitation of facts and a blind carrying out of procedures" (p. 52) that creates a robotic response in people who think that if they cannot answer the question immediately they will never get it. Paulos (1991) describes the overabundance of school computation as "boring, tiresome, and oppressive" and blames school mathematics for coloring society's understanding of real mathematics (p. 52). In fact, Paulos (1991) makes the argument that what gets taught in schools tends not to be real mathematics at all:

> Imagine that 90 per cent of every course in English up until college was devoted to grammar and the diagramming of sentences. Would graduates have any feeling for literature? Or consider a conservatory, which devotes 90 per cent of its efforts to the practicing of the scales. Would its students develop an appreciation or understanding of music? The answer, of course, is no, but that, given proper allowance for hyperbole, is what frequently happens in our mathematics classes. (p. 52)

Mathematician Peter Taylor, who writes frequently about school mathematics, argues that it is the very importance of mathematics in schools that has destroyed it. In a speech he gave to the International Conference of Mathematics Educators, Taylor stated:

> Mathematics is considered important—too important to leave anything to chance or to the whims of the individual teacher, so everything must be detailed. (The irony there is that it's the importance of the subject which has killed it—don't miss that one. Music, when it is taught, which is rare, can be taught with complete freedom only because it is considered unimportant.) And secondly, mathematics is a linear subject and it is well known that you can't learn one thing until you've learned what comes before, so you can't afford to leave anything out or the students will be crippled down the line. And thirdly, all these things that you have to know are not framed in terms of method because method is hard to get hold of, cannot be easily described, rather they are catalogued in terms of fragments of technique. (Taylor & Sinclair, 2000, p. 3)

Taylor (1997) wrote a piece that appeared in the *Globe and Mail* about the sorry state of school mathematics. The opinion piece stemmed from his frustration with his daughter's high school mathematics experience. Falling in line with Paulos, Taylor (1997) refers to the tediousness of the high school mathematics curriculum that does not reflect in any way the nature of mathematics in the slightest, and he writes, "Something is wrong when I'm a mathematician and

I can't read my daughter's math text for pleasure. It's not forging 'elite skills' in anyone—and the problem isn't money." The term 'elite skills' refers to a *Globe and Mail* headline about young Canadians lacking elite skills. Taylor shoulders part of the blame, as a mathematician, for allowing his discipline to be distorted.

Taylor and Paulos are articulating the state of affairs in mathematics education. Their dissatisfaction can be traced partly to historical characters such as Pythagoras and Descartes, but they are announcing that the Mathematics Emperor "has no clothes" and they are trying to get a "reconception" of the discipline out into a public place. Importantly, their deep understanding of the subject and what mathematics really is about and what mathematicians really do is being brought into public view. The secret is out. Sort of.

Reconsidering Mathematics in Schools

If Taylor and Paulos are representative of reconceptualist thinking, what is being done about the nature of mathematics in schools? Mathematics education research is responding to this complex issue, but it is still a relatively new field (Boaler, 2002). Regarding mathematics education research, Sierpinska (2004) made a strong declaration during a presentation to the Canadian Mathematics Education Study Group when she stated:

> We need a lot more critical studies that would question and find weak points in our research methodologies, theories, as well as conclusions and other products of research. We need, indeed, to be "stirred up"—cognitively stimulated in mathematics education research. There should be more debate, more sharp criticism, more hard analytic thinking about the phenomena we study and about the validity of our claims about them. (p. 22)

While Taylor is trying to "stir up" the view of mathematics in the public forum, there is a body of scholarly mathematics education research that is beginning to look deeply at how the nature of mathematics might be reconsidered (Friesen, 2000). Unlike the reconceptualist movement in curriculum theory and development, there is yet no clearly defined school of reconceptualists in mathematics education. When searching through scholarly journals and texts about mathematics education, there is an abundance of research that stems from a psychological framework. There is a privileging of research that comes from quantitative studies of how children learn mathematics, studies on mathematical achievement, studies regarding child development and mathematics, and studies on teaching approaches for mathematics. While this research can be helpful for those looking for empirical data, on gender and achievement, for example, it is the work of people who think and write about mathematics as a human, living enterprise that I find the most intriguing (Applebaum, 2003; Davis, 1995, 1996; Ernest, 1996, 2003; Friesen, 2000, 2008; Friesen, Clifford, & Jardine, 2003; Gordon Calvert, 2001; Jardine, 1994, 1998; Jardine, Clifford, & Friesen, 2003; Jardine, Friesen, & Clifford, 2003; King, 1992; Lakoff & Núñez, 2000; Sinclair, 2001). Such researchers 'stir up' notions about the nature of mathematics itself.

As a teacher of pre-service elementary teachers taking up the teaching of mathematics, as someone who has worked in schools with young children, and as a former mentor for classroom teachers, I can state the traditional view of mathematics is very much alive and living in classrooms. Much of my energy is now occupied by trying to disrupt the traditional view of mathematics.

Mathematics education research that questions and challenges the nature of mathematics and its connection to human beings is what I find myself pulled toward. It is the most helpful in allowing me to further theorize about mathematics education in the lives of children and pre-service teachers. But to be able to disrupt the traditional view of school mathematics means that it is necessary for me to better understand the nature of mathematics as described by those who wish to 'stir things up.' This means an unpacking of my own mathematical experiences and prejudices and laying bare my own teaching of the discipline. After all, I was immersed in the fishbowl of the absolutism of mathematics and have been forced by my work to look past the bounded glass to other positions and possibilities.

To come to understand the character of mathematics as a living discipline I have turned to the work of such scholars as Peter Applebaum, Brent Davis, Paul Ernest, Sharon Friesen, Lynn Gordon Calvert, David Jardine, George Lakoff and Rafael Núñez, John Allen Paulos, Natalie Sinclair, Peter Taylor, Jennifer Thom, and Jo Towers. These scholars have unearthed for me a way into mathematics so that I have come to understand it as a human dialogic living enterprise complete with aesthetic dimensions. Many of the scholars listed above, as well as many other Canadian mathematics education scholars not mentioned in this book, have done much "to illuminate mathematical thinking and the mathematics brought forth by systemic and dynamic phenomena" (Thom, 2003, p. 187). For the purposes of this chapter, I wish to focus on those who write about the nature of mathematics in particular. It is this research that lays the foundational challenges to the status quo. This body of research allows me to move across the bridge of reconception. It is the work of the reconceptualists that has brought to light my own difficulty in working with children and pre-service teachers around curricular issues in mathematics. Their writing is not only providing possibilities for a way forward, but also helping me move across the divide of personal indoctrination away from the secret society of Pythagoras and Descartes.

Mathematics with its "rich topography of relations" (Jardine, Clifford, & Friesen, 2008, p. 5) is a human endeavor. Thought of as a living system, mathematics has a strong ecological nature (Davis, 1995, 1996; Friesen, 2000; Friesen, Clifford, & Jardine, 2003; Gordon Calvert, 2001; Jardine, 1994, 1998; Jardine, Clifford, and Friesen, 2003, 2008; Jardine, Friesen, & Clifford, 2003) made up of interconnections and interrelationships. As Davis (1996) writes, "When we speak of ecology, then, we speak of everything that shapes our being" (p. 58).

It is not preexistent; it does not live in any one of us, "yet it requires us" (Friesen, 2000, p. 10). It can be overwhelming to think about the ecological nature of mathematics. Even as a body of knowledge, Gordon Calvert (2001) describes its fractal nature made up of "evolving networks on many levels" (p. 36) as an image of a dynamic living system:

> From each level, we can see the body of mathematics on different scales—as an interconnected set of problems or human concerns. On a larger scale we can imagine that geometry, calculus, and algebra form major points of intersection within the networked body of mathematics. We can then select and bring into focus any one of these points, such as geometry. Within geometry another network arises, including points for Euclidean geometry, fractal geometry, and topology. (p. 36)

Recovering the connectedness of mathematics to the world and to the body, an idea that was severed courtesy of Descartes and Newton, is critical for coming to understand the living nature of mathematics. "Mathematics is truly about us and our world" (Davis, 1996, p. 81). Jardine (1998) puts it this way:

> Making mathematics seem more human entails the darkness of humus and earthliness, with all its interweaving and intersecting threads. But it also entails a sense of something out of which things can grow, something alive or sustaining of life, something generative. (p. 66)

No longer seen as an isolated discipline, mathematics requires us, and Earth, for its existence. "Mathematics must become earthen in how it is understood, how it is taught, and how it is 'grounded'" (Jardine, 1998, p. 75). Drawing from Wittgenstein, Jardine (1998) writes, "We cannot give it a boundary that could prevent it from intertwining with our lives and the life of the Earth.... No matter how careful we are in our drawing of boundaries, mathematics interweaves with the fabric of the Earth" (p. 75).

Traditional school mathematics is deeply severed from Earth and from the bodies of children in elementary school classrooms. Boundaries are drawn around the isolated subject to contain it (Jardine, 1998). "We endured [school mathematics] because we had no other choice" (King, 1992, p. 16). But the nature of mathematics that I have come to understand is grounded in a bodily experience of the world. "Mathematics is embodied mathematics" (Lakoff & Núñez, 2000, p. 347).

There are features to the territory of mathematics as a living discipline. It has an ontological, unfinished nature filled with partial understandings and uncertainties kept alive through conversation (Davis, 1996; Ernest, 2003; Friesen, 2000; Gordon Calvert, 2001; Jardine, 1994, 1998; Jardine, Clifford, & Friesen, 2003, Jardine, Friesen, & Clifford, 2003).

Our partial understandings in learning the ways of this territory are a fea-ture of the place. They don't belong solely to a person. Anything that is living is not wholly worked out, it is not complete, it is always partial. When you are standing in mathematics, you are standing in a living place. It's moving. It's alive (Friesen, 2000, p. 79). Standing in a living place means you are stand-ing in a terrain filled with ancestry and tracings of centuries of mathematics. "Being part of a living discipline means that you are dropped into a conversa-tion that is centuries old" (Friesen, 2000, p. 94).

The history of mathematics is gathered and held within itself. When children gather to work out the Pythagorean theorem, it is the same 3-4-5 that Pythagoras worked with. Jardine, Friesen, and Clifford (2003) draw from Gadamer, and they write, "Each thing is all the codependent arisings that brought it here, and to understand this particular thing is to understand its standing in an 'inheritance that it belongs to'" (p. 43). Recently, in talking about the nature of mathematics and how beautiful it can be with a Language Arts teacher friend of mine, she remarked, "Well, I just don't get it. 4 + 4 = 8 doesn't seem beautiful to me." Standing as a math fact, 4 + 4 = 8 bears no tracings of how it came to be, and it holds no memory (Friesen, 2000). If you forget that 4 + 4 = 8, then you are lost. "And even if you simply memorize and remember this fact, you have no way to go on, since it also carries no memory or trace of directionality and place" (Friesen, Clifford, & Jardine, 1998, p. 11). Unfortu-nately, elementary classrooms are often spilling over with math facts that hold no memory. Viewed in this way, it is no wonder that it becomes difficult to care for a topic that seems so sterile and superficial. Thus, it gets taught as sterile:

> Mathematics becomes akin to a tourist attraction, something to look at but never enter into, open up, and learn to live with. And we, in turn, become akin to curricular tourists ready to be momentarily entertained and amused. However, since we just see the thin, tarted-up, presentable surface of things, we, along with our children, become equally subject to boredom [and] frustration. (Jardine, 1994, p. 265)

But taken up with all its complexity, mathematics joins us to a world of rhythm and pattern (Gordon Calvert, 2001, p. 52) that may contain great beauty. As human beings we are good at sensing balance, motion, and symmetry (Lakoff & Núñez, 2000), yet, as King (1992) points out:

> The fact that mathematics presently lies outside the artistic range of most people is the fault of neither the audience nor of mathematics. What has gone wrong is the manner of presentation. How else can there exist a person who likes poetry and hates mathematics? Properly presented, they are much the same. (p. 277)

According to King, the aesthetic value of mathematics is "as clearly defined as that of music or poetry" (1992, p. 23). But often in schools mathematics gets linked to music or art as a way to hook the students' interest. Nathalie Sinclair (2001) argues that the aesthetic flow should and can originate in mathematics:

> Appealing to students' love of something else … in the hopes of increasing their interest level in mathematics is that it [is problematic in that it] endorses the belief that mathematics itself is an aesthetically sterile domain, or at least whose potentialities are only realized through engagement with external domains of interest. Such linking also tends to undermine the aesthetic, expressive and transformative possibilities of mathematics itself. (p. 25)

When I first discovered the existence of the Fibonacci sequence it changed me. No longer could I look at sunflowers and daisies without an incredible sense of awe and wonder. Before, they were just flowers that had petals and seeds. They were pretty flowers but hardly phenomenal. Why would I pay attention to how many seeds or petals they contained? But when I learned that these objects contained Fibonacci numbers of petals and seeds I was dumbfounded. I had found joy in this discovery:

> The joy of mathematics is similar to the experience of discovering something for the first time. It is an almost child-like feeling of wonder. Once you have experienced it

you will not forget that feeling—it can be as exciting as looking into a microscope for the first time and seeing things that have always been around you that you have been unable to see before. (Pappas, 1989, Introduction)

My deepened understanding of the nature of mathematics led me to pay attention, demanded I pay attention, to the world around me and see for the first time what was being presented to me.

To see the artfulness in mathematics, Davis (1996) also points to such mathematical formulations as the Fibonacci sequence

> through which it is possible to link our experiences of phenomena as diverse as spirals and sunflowers to music, hinting at a mysterious "pattern which connects" (to borrow George Bateson's phrase) not only the seen to the heard but the singular to the cyclical and the part to the whole. (p. 156)

The area of fractal geometry—pictures of nature, for a simplistic definition—allows learners access to complex and beautiful shapes as well as access to sophisticated mathematics that can be appreciated during a walk in any woods filled with ferns, trees, and flowers.

Objects that are balanced and symmetrical are aesthetically pleasing. As human beings we are good at sensing balance, motion, and symmetry (Lakoff & Núñez, 2000), and therefore basic human activities "such as artistic expression and construction" hold firm the "relationship of mathematics to basic human sensitivities such as rhythm and fit" (Sinclair, 2001, p. 26). Friesen et al. (2003) describe a young girl working on a spiral consisting of right-angle triangles. The care she gives her work and the way she proceeds is art-full:

> Placing her ruler across the two points that she has calculated and measured, she ever so carefully draws the first light pencil line. Then checking to ensure the accuracy of the line, Jenny draws the second, now darker line over the first line. She removes the ruler from the paper and critically analyzes her work. "Good, it's good," she seems to say. And then she repeats the process, recursively adding the next and then the next line to the geometrical drawing. Sometimes a smile of intense satisfaction crosses her face. Sometimes fellow students come by to inquire about her work. "Wow Jenny, that is so beautiful," they say as they admire the emerging form. Jenny smiles and then goes back to calculating, measuring and drawing. Each line is precise. Each calculation is exact. (Friesen et al., 2003, p. 2)

To take up mathematics in such a way, with wonder and awe for the simplicity and beauty of such a thing as the Pythagorean theorem, means acknowledging the living, unfinished, temporal, dialogic, aesthetic nature of mathematics. When we enter the territory of such a place we no longer can remain the same. Our understanding—of the territory and of ourselves—changes, as have I, in working through the scholarly work of the reconceptualists and putting into language what I have come to understand. All understanding is self-understanding, according to Gadamer (1989), and my work thus far on the nature of mathematics has shown this to be true. It is no wonder that the reconceptualists take an interpretive stance in their work. In fact, it only makes sense. The more I continue to explore the nature of mathematics, the more I realize that for me to ask the question "What is mathematics?" suggests that I believe "that we can somehow consider the body of knowledge as determinable, fixable, and separable from ourselves—as though we could somehow step outside of our mathematics" (Davis, 1996, p. 80). Perhaps the question that should be asked is not "What is mathematics?" but rather "What are we that we may know mathematics" (Davis, 1996, p. 81)?

The reconceptualists have helped bring me across the bridge to a new place. They have helped put into perspective my own mathematical education, made me aware of the place where many of my education students dwell, and have allowed me to focus on the mathematical understanding of teachers of young children. Now my task is to reveal the secret nature of mathematics to education students and recover it, as Davis suggests at the beginning of this chapter, as a humanity, as a living human enterprise. It is time for the secret to be out for good.

· 4 ·

HERMENEUTICS:
AN ONTOLOGICAL TURN

Introduction

I first came across the term 'hermeneutics' eight years ago during a chat with a researcher who was spending time at the school where I was teaching grade one. Making small talk, I remember asking the researcher what type of research he did, and he said, "Hermeneutics." It was completely foreign to me. I confessed my ignorance and he went on to say something like, "It is concerned with what happens in the space between this and this." As he spoke he motioned with his hands in the air. I found his response fascinating. Children began to pour into the classroom and all I was left with was this one sentence. My curiosity led me to seek out people and literature that could open up the word 'hermeneutics' for me. After my first graduate class on the topic, I felt like a whole world had cracked open, but I found the subject so slippery. Throughout the semester I tried to explain it to others who knew I was taking the graduate course. I would be asked, "What is this herma-hermen- what do you call it?" And I would find myself somewhat tongue-tied in response. I would say things such as the following:

"Hermeneutics is about interpretation of text."
"It's about conversation."

"Hermeneutics is about human understanding."
"It is about the cultivation of character."
"It's a type of interpretive inquiry."
"Hermeneutics is about transformation."
"It's about remembering and forgetting."
"It's about possibility."
"Hermeneutics is concerned with what gets concealed and what gets revealed."

These expressions sounded good. When pressed further, I would stumble and bring in words such as 'ancestry,' 'tradition,' 'historical consciousness,' and 'language' but I really did not understand what it was I was saying. Since that first course on Hans G. Gadamer's hermeneutics, I have figured out why I fumbled so much with the language. Hermeneutics is difficult. The more time I spend reading about and discussing hermeneutics, the more it spreads out. I know that we are always on the way to understanding, and never can arrive. However, I do feel as though I am beginning to 'know my way around' the terrain. Hermeneutics provides a way forward with the topic of mathematics and a "returning to the original difficulty" (Caputo, 1987, p. 7).

Hermeneutical Tracings

From Aristotle to Husserl

An etymological tracing of the word 'hermeneutics' leads to Hermes. A messenger god, Hermes delivered messages between the gods and between mortals and the gods. Considered also as a trickster and thief, he is associated with borders and boundaries, with youthfulness, fertility, and impudence (Caputo, 1987; Davis, 1996; Gadamer, 1989; Jardine, 1998; Jardine, Clifford, & Friesen, 2003, 2008; R. Palmer, 1980; Smith, 1994).

The history of hermeneutics goes back as far as Aristotle (Smith, 1994). In the fifteenth century it emerged as a way of interpreting biblical texts. Smith notes the important contribution of nineteenth-century thinker Friedrich Schleiermacher, who distinguished three themes in hermeneutic inquiry, "the inherent creativity of interpretation, the pivotal role of language in human understanding, and the interplay of part and whole in the process of interpretation" (Smith, 1991, p. 190). Schleiermacher's biographer, Wilhelm Dilthey, developed a more historically conscious methodological hermeneutics. Dilthey was "concerned to legitimate the work of the human sciences epistemologically. He was dominated throughout by the questions of what is truly given" (Gadamer, 1989, p. 64). He sought to explore "understanding" as emerging from "lived experience," *Erlebnis* (Smith, 1994). Dilthey was influenced by the work of Edmund Husserl, who developed the notion of the "life-world." For Husserl, the intentionality of experience was extremely important. Every experience is an experience of something and not an experience in general:

> Through his theory of intentionality Husserl showed that we never think or interpret "in general" as a rhetorical activity that bears no necessary connection to the world at large. Rather, thinking and interpreting are always and everywhere precisely about the world. (Smith, 1994, p. 108)

Husserl and Intentionality

The notion of intentionality was very important to the evolution of phenomenology and hermeneutics. Husserl was the father of phenomenology, and his work was an interruption of the subjectivization of experience. Phenomenology would turn "to the things themselves" in its felt immediacy. Not concerned with explaining the world, phenomenology would seek to describe "as closely as possible the way the world makes itself evident to awareness, the way things first arise in our direct, sensorial experience" (Abram, 1996, p. 35). There could not be a split between subjectivity and objectivity because "my subjectivity gets its bearings from the very world that I take as my object" according to Smith (1994, p. 108). Husserl believed describing the essence of something to be the goal of phenomenology. However, Husserl's student, Martin Heidegger, would call Husserl's notion of essence into question. To search for essence assumed that the life-world is a static world.

Heidegger's Ontological Shift

Heidegger's work brought a major shift to hermeneutics—an ontological shift: "Interpretation is the primordial condition of human self-understanding so that a phenomenology of Being reveals its fundamental mode to be precisely hermeneutical" (Smith, 1994, p. 109). Heidegger's work cast a shadow on Dilthey's project of a method for the human sciences because "it [the method] bore the same character and quality as that to which it sought access" (Smith, 1994, p. 109).

Gadamer's Ontology of Understanding

This brings me to the philosopher whose work speaks to me most strongly. Hans G. Gadamer's hermeneutics in *Truth and Method* follows the lead of Heidegger in making hermeneutics an ontology of understanding. A student of Heidegger, Gadamer took on the notion of method in the human sciences. He argued that the appropriate method for interpreting any phenomena "could only be disclosed by the phenomena itself through a kind of Socratic dialogical engagement between question and phenomenon" (Smith, 1994, p. 110). Through his work, Gadamer looked to uncover the nature of human understanding. He argued that truth and method were at odds with each other. He was critical of two approaches to the human sciences. He argued against approaches to the humanities that were based on the natural sciences and he challenged the traditional approach stemming from Dilthey that involved searching for a correct interpretation of text that looked to the original author's meaning. He maintained that human beings have a historically effected consciousness and are embedded in the history and culture that shape us. Caputo (1987) takes on hermeneutics and its connection with deconstruction. I appreciate his suspicion of the "easy way out" (1987, p. 3), as well as Caputo's concept of attending to the ruptures in existence before the metaphysics of presence smoothes the irregularities over. Caputo writes, "Hermeneutics always has to do with keeping the difficulty of life alive and with keeping its distance from the easy assurances of metaphysics" (1987, p. 3). While Gadamer captures my hermeneutic attention most fully, I will attend to Caputo's suspicion of the easy way out and his idea of returning to the original difficulty in this scholarship.

Choosing the Method: Did I Have a Choice?

My real concern was and is philosophic: not what we do or what we ought to do, but what happens to us over and above our wanting and doing. (Gadamer, 1989, p. xxviii)

I have not selected the topic of mathematics and then decided to do an interpretation of it. I did not decide that I would take up hermeneutics and then search for a topic to interpret. As Jardine (2008) writes, "Interpretation does not begin with me" (p. 199). It begins when something addresses us, when something strikes us. When my elementary pre-service teachers responded to the teaching of mathematics with such disdain and frustration in that first class in January 2004, I was thrown by it. I could not ignore it, or pretend it did not happen. In fact, I had begun my doctoral program with intentions of researching humanity and technology, but I was captured by the problem mathematics was presenting to student teachers and it would not let me go. Sitting in that classroom before me were educated people who desired to become elementary school teachers; they had succeeded in school mathematics for the most part and were able to survive in a mathematical world. Yet they displayed incredible anxiety and unease with teaching mathematics. What was going on? I turned to face the event. A world began to open up to me and for me around mathematics and its connection to young children and student teachers. Every classroom I walked into as a field advisor I saw things differently. The difficulties surrounding mathematics and student teaching were presenting themselves at every turn. "Interpretation thus 'finds itself' in its topic" (Jardine, 2008, p. 199). It would be naïve of me to think I will come to an absolute solution to the teaching and learning of mathematics, particularly as it intertwines in the lives of student teachers and young children. It was during the first university mathematics education class in January 2004 that I was hit with the difficulty—the difficulties of teaching, the difficulties of mathematics, the difficulties of living a good life with children on Earth right now. From unearthing the nature of mathematics I understand that difficulty is at the heart of a discipline that is living (Clifford & Friesen, 2008; Friesen, 2000). This book takes up the topic of mathematics teaching and children and I will make the difficult liminal experience readable and understandable and decipherable as something more than simply an array of problems to be fixed. More strongly put, this work suggests that there are deep and irremediable difficulties inherent in the liminal space traversed by student teachers that cannot and should not be fixed. Not all difficulties are a result of lack of effort or diligence or preparation or information, and to understand student teaching in such a pathological fashion forgoes a collective, communal understanding of its deep human meaning, with all of the ways in which such commonly held understanding can help to make the burden of those difficulties more bearable. (Jardine, 1998, p. 124)

I am not interested, as Jardine points out, to "fix" the problems with mathematics in schools. To believe this is possible would mean that I do not really understand the living nature of mathematics or schools or teaching in the first place. What is hoped is that I can come to better understand mathematics and its relations to student teachers and children by exploring and excavating the topography.

I Am in the World

> History does not belong to us: we belong to it. Long before we understand ourselves through the process of self-examination, we understand ourselves in a self evident way in the family, society, and state in which we live. The focus of subjectivity is a distorting mirror. (Gadamer, 1989, pp. 276–277)

In reading Gadamer's words, I am reminded yet again of my own subjectivity. Until I began reading Gadamer's work, I thought of prejudice strictly as a negative attribute. However, from Gadamer's perspective, prejudice does not block access to the world. It is our access to the world. I have been a teacher for more than twenty years in four different provinces. I have worked in rural, suburban, and urban settings. All that I have experienced and lived through as a woman and a student and a teacher of mathematics is a part of who I am, what I see, and what I hear. As I come to better understand teaching mathematics, I come to better understand myself. Who I am as the author is a very necessary part of the writing. I cannot separate myself from the topic. I do not want to "translate my own subjectivity out of the picture but to take it up with a new sense of responsibility" (Smith, 1994, p. 127). While I can never be fully conscious of all of my prejudices,

> it is important to acknowledge their existence and recognize the necessary constraints they place on our knowing and doing. Through our prejudices we are able to select or make distinctions on ideas and objects, but for every distinction made in one way, we close off possibilities for other distinctions to be made. (Gordon Calvert, 2001, p. 53)

When I delve into the notion of insiders and outsiders around mathematics, I lay bare my own experiences of being perceived an outsider and an insider. As I explore the idea of a cover version within mathematics, I expose my own implications in such a reading of school mathematics. The writing I have done to this point of the book is rife with prejudice, such as my view of the nature of mathematics as a living discipline, how I understand hermeneutics and its connections to my topic, and why I am doing this work in the first place. Perhaps Smith is on to something when he states, "All writing is autobiographical" (1994, p. 127), for indeed I cannot be separated from or translated out of the topic. However, it is critical that I make clear that I am not the topic. This will not be 'my story.'

Dialogical Nature of Hermeneutics

As an ontology of understanding, hermeneutics "avoids both the subjec-
tivizing involved in making interpretation a psychological process, and an
objectivizing which omits/denies the interpretive moment in the reader"
(R. Palmer, 1999). Readers of my text will be able to share my experiences.
I am not looking to pass on information to the reader through my text; I hope
to evoke in them a new way of understanding the teaching of mathematics
to children and a new understanding of who they might be in the world.
I hope to create a dialogical text that not only creates meaning for the readers
but also contains a critical reflexivity about my own pedagogic actions. In
doing this, I am attempting to avoid any sense of romanticism around
teaching young children mathematics or working with student teachers as
they work with young children. Being critically reflexive in the text is the
responsibility that I have as the author.

Mathematics Requires a World

Just as I am in the world, mathematics, too, requires a world (Davis,
1996; Friesen, 2000; Jardine, 1998, 2006). As a living enterprise it has an
ancestry:

> Mathematics is something that lives in its "being handed along" (Gadamer 1989, 284).
> It is something we inherit. We can't individually construct it nor can we passively
> submit to it. Mathematics comes to us through the generations of mathematicians that
> explored its contours, created its ways, and mapped its paths. (Friesen, 2000, p. 77)

Simply put, things are all their relations. Mathematics, with its ontological
character of being unfinished, of not yet being itself completely, is filled with
partial understandings. "The reality of something like mathematics is thus
'standing in a horizon of … still undecided future possibilities' (Gadamer,
1989, p. 112)" (Jardine, Clifford, & Friesen, 2003, p. 131). Gadamer (1989)
writes specifically about mathematics and its connection with humanity when
he states, "The history of mathematics … is also a part of the history of the
human spirit and reflects its destinies" (p. 283). Unfortunately, the ancestors of
mathematics do not seem to be invited to show themselves in many elementary
classrooms. Instead, mathematics stands severed from its ancestry with a sort

of amnesia. Davis and Hersh (1986) write about the importance of temporality and mathematics:

> A de-temporalized mathematics cannot tell us what mathematics is, why mathematics is true, why it is beautiful, how it came to be, or why anybody should care a fig about it. But if one places mathematics squarely within human time and experience, it becomes a warm and rich source of possible meanings and action. Its ultimate mystery is never dispelled, yet it is exhibited as one of the prime creations of the human intellect. (p. 201)

Like us, mathematics has a past, a present, and a yet to be determined future of possibilities. Hermeneutics is marked by finitude. Understanding is always on the way. "This understanding of our temporal nature Gadamer called 'effective historical consciousness' and its character is revealed most pristinely in the structure and function of language" (Smith, 1994, p. 111).

Memory

Memory is an important aspect of understanding. Hermeneutics does not have a method—it has a memory. Memory is a form of knowing and being in the world. You do not simply seize memories; you live them and they shape you. Gadamer (1989) states that memory must be formed. He also wants to rescue memory as a psychological faculty and to have it be seen as "an essential element of the finite historical being of man [sic]" (p. 16). Chapter 1 pointed to moments of my own forgetfulness in the pedagogic sense. Gadamer (1989) writes that forgetfulness is necessary for remembering. "Only by forgetting does the mind have the possibility of total renewal, the capacity to see everything with fresh eyes, so that what is long familiar fuses with the new into a many leveled unity" (p. 16).

Truth

Alethia, the Greek word for 'truth,' comes from *Lethe*, which is Greek for 'forgetfulness.' In Greek mythology, Lethe was a river in Hades and those who drank from it would suffer forgetfulness. According to Greek mythology, souls who were to be reincarnated were to drink from the river Lethe so they would not remember their past lives. Therefore, alethia also means remembering what has been forgotten, to enliven that which used to be dead. Hermeneutically speaking, it also means un-concealment; as Jardine and

Friesen (2008) state, "It is the word hermeneutics invokes for truth as uncovering, opening up" (p. 133). But truth in the hermeneutical sense does not mean coming to a single unambiguous truth. Rather, "the interpretive truth of this tale lies in whether it can be read in a way that might help us more openly and generously understand the lives we are already living" (Jardine, 2008, p. 200). The truth of my interpretation of the difficulties of mathematics in schools for teachers and children will not lie in the story I tell, but whether or not what I write will be telling of something. "Truth is not a given state of affairs, but is always yet-to-be-decided by this reader's or that reader's giving" (Jardine, 2008, p. 200) of what has been written. For readers to discover alethia in what I write, I need to reveal what has been concealed in my topic; for example, by bringing to light the notion of a cover version in mathematics, in which students can pass through grade school, high school, and university with a certain functional understanding of mathematics and then become elementary teachers who recapitulate the same version of mathematics to other human beings, is something that I believe needs to be revealed and peeled back, so together, the readers and I can see what might lie at its center. In a conversation with the text, the reader must try to remain open to the meaning of the text for the "multiplicity of possibilities" (Gadamer, 1989, p. 268).

> If a person fails to hear what the other person is really saying, he will not be able to fit what he has misunderstood into the range of his own various expectations of meaning.... The hermeneutic task becomes of itself a questioning of things and is always in part so defined.... This places hermeneutic work on a firm basis. (Gadamer, 1989, p. 269)

What is necessary for engaging in a conversation with a text or a person is to read or listen for the truth in what the other is saying. I need to ask, "What is this text (or person or situation) asking of me?"

Understanding through Conversation

> Conversation is a process of coming to an understanding. (Gadamer, 1989, p. 385)

Conversation is central to Gadamer's hermeneutics. It is about something, not just idle talk or chitchat. Conversation takes place through language and, therefore, understanding is linguistically mediated. In a true conversation opinions do not come into play. In fact, if you get to the point at which participants are stating their opinion, the conversation, if there was one, is already over. In a conversation, each member is trying to understand what is being asked of him or her, and to consider what the other is saying as being true of something. Through conversation, new understandings are reached and new meanings made. In coming to an agreement through conversation, a common language has been found. In dialogue we uncover our conflicting prejudices and dissolve our subjectivity. Instead of holding fast to our position, we listen to the other, and remain open to the other, whose horizon may be different from our own (Davis, 1996).

Horizons

'Fusion of Horizons' is a term that Gadamer (1989) uses to describe what happens in a real conversation. Gadamer (1989) takes the meaning of horizon—"the range of vision that includes everything that can be seen from a particular vantage point" (p. 302) and applies it to human thinking:

> A person who has no horizon does not see far enough and hence over-values what is nearest to him. On the other hand, "to have a horizon" means not being limited to what is nearby but being able to see beyond it. A person who has an [sic] horizon knows the relative significance of everything within this horizon, whether it is near or far, great or small. Similarly, working out the hermeneutical situation means acquiring the right horizon of inquiry for the questions evoked by the encounter with tradition. (p. 302)

Our horizons of the present are not fixed but are continually being formed because we continually test our prejudices (Gadamer, 1989). "An important part of this testing occurs in encountering the past and in understanding the tradition from which we come. Hence, the horizon of the present cannot be formed without the past" (Gadamer, 1989, p. 307). The old and the new are constantly being fused (Gadamer, 1989). "A fusion of horizons is an event of truth" (Davis, 1996, p. 27). The degree of awareness of the participants in a fusion of horizons is intriguing. Davis (1996) describes it like this:

> We can never be fully aware that a conversation is taking place. We can, however, be aware that one has taken place. When understandings have changed, when a new commonsense has been established—when self and other have been altered—it has happened. (p. 29)

There is a sense of self-forgetfulness in the midst of a conversation:

> When one is engaged in a good conversation, there is a certain quality of self forgetfulness as one gives oneself over to the conversation itself, so that the truth

realized in the conversation is never the possession of any one of the speakers or camps, but rather it is something that all concerned realize they share together. (Smith, 1994, p. 120)

Conversation with Mathematics

Conversation does not exclude mathematics. Gadamer (1989) writes: "To conduct a conversation means to allow oneself to be conducted by the subject matter to which the partners in the dialogue are oriented" (p. 367). Jardine (1998) puts it this way: "Interpretive work inevitably begins with a living subject [mathematics] in a living dialogue with the life that surrounds us" (p. 50). Drawing from Gadamer, Jardine (1998) continues:

> To reach an understanding … is not merely a matter of putting oneself forward and successfully asserting one's own point of view, but being transformed into a communion in which we do not remain what we were (Gadamer, 1989, p. 379). In such a case, interpretive work is profoundly pedagogic. (p. 50)

Friesen (2000) brings mathematics into the fray, and she writes, "In conversations the participants do not attempt to control the subject matter, but rather are deeply engaged in attempting to understand the issue at hand. The subject (mathematics) participates in the conversation—mathematics speaks" (p. 81).

Many of the reconceptualists in mathematics education take up conversation as a way of understanding mathematics. Gordon Calvert's (2001) text, *Mathematical Conversations with the Practice of Mathematics*, is but one example of research that takes up the teaching of mathematics through conversation. Such work demonstrates the significance of conversation in the pedagogic relationship.

Understanding the Part and the Whole

The hermeneutic circle is ubiquitous in descriptions of hermeneutics. Like Heidegger, Gadamer took up Schleiermacher's notion of the interplay between the part and the whole in the process of interpretation (Smith, 1994). This "back and forth movement between the particular and the general, is more popularly referred to as the 'hermeneutic circle'" (Davis, 1996, p. 21). As I move from the specific to the general, my understanding of both becomes deepened and this affects all other understandings. Gadamer brings the interpretive consciousness into his articulation of the hermeneutic circle.

The hermeneutic inquirer cannot be a detached observer, as Davis (1996) points out:

> Rather, the interpreter recognizes his or her complicity in shaping the phenomenon, simultaneously affecting and affected by both the particular and the general, thus wholly embedded in the situation. In other words, the "object" of the hermeneutic inquiry is a moving target. (p. 22)

By writing about particular pedagogic moments of awareness with students and mathematics, I come to understand not only the particular moment, but I come to a different understanding of the nature of mathematics in general, of teaching in general, and of difficulties of living on Earth ethically with one another. By remaining open in this cyclical process, I am in a state of becoming.

Play and Pedagogy

As a pedagogue, the idea of play in hermeneutics is intriguing. In Greek, 'play' (*paidia*) and 'education' (*paideia*) are similar: "Both terms arise from an original reference to the activity of the child (pais), an echo of which can be heard in the word 'pedagogy' (paidagogos)" (Davis 1996, p. 212). When it comes to the element of 'play' in *Truth and Method* (1989) "metaphorical usage has methodological priority" (p. 103). Concentrating on play as a metaphor, Gadamer (1989) playfully takes it up: "We find talk of the play of light, the play of waves, the play of gears or parts of machinery, the interplay of limbs, the play of forces, the play of gnats, even a play on words" (p. 103). In many classrooms there seems little space for play with mathematics. Friesen (2000) argues that play is exactly what is needed for the creation of new mathematics:

> Play, understood as something frivolous, opposed to work, off-task behavior, is not welcomed into most mathematics classrooms. But play is exactly what is needed. It is only play that can entice us to the type of repetition and rehearsal that is needed to learn how to inhabit the mathematical landscape and how to create new mathematics. (p. 101)

Supporting Friesen's argument, Davis (1996) states that play "is a phenomenon that has tended to be shallowly understood and, in consequence, almost universally scorned by mathematics teachers" (p. 214). Gadamer refers to play as a form or renewal, with no goal in the movement of play that "brings it to an end; rather, it renews itself in constant repetition" (1989, p. 103). As an element of hermeneutics, I will attempt, for example, to write about what is 'at play' in the notion of student teachers having a cover version of mathematics. I wish to uncover the pedagogic connection, as well as the severance, of play that exists in elementary mathematics classrooms.

Possibilities

Perhaps the reason I feel a strong affinity for Gadamer's hermeneutics and the reason he advances my work is because of what I understand as a hermeneutics of possibility, of hope, of generosity, and of responsibility. Engaging in conversation and looking for what might be true in what the other has to offer is a truly generous act. Gadamer describes the task of hermeneutics in the foreword to the second edition of *Truth and Method*. He writes, "What man [sic] needs is not just the persistent posing of ultimate questions, but the sense of what is feasible, what is possible, what is correct, here and now" (p. xxxviii). Hermes may appear when I least expect it. One needs to be ready for that possibility and remain open to what presents itself. We must be prepared to deepen self-understanding if we are to reach any hermeneutical truth. That is our responsibility as ethical human beings. There is a future for children and teachers at stake.

· 5 ·

ANOTHER WAY OF BEING WITH CHILDREN AND MATHEMATICS

The Question Is Ontological

I have spent time in many classrooms. I have had numerous conversations with other teachers. I have been witness to children's learning. What I notice is this: As teachers we are often blinded by our practice in the immediacy of events occurring in the very classrooms in which we converse with learners and implement curricula. As well, we tend to be most concerned with the following two questions—that is, two questions reflecting the perceived business of the day:

1. What am I doing?
2. How am I going to do it?

Both these questions are riddled with a metaphysics of presence—is it any wonder what seems to matter most to us as teachers is the 'here' and 'now'! However, as a result of this reflexive hermeneutic endeavor, I have been led to wonder: Are we not forgetting the question that is of utmost importance? Are we forgetting to challenge ourselves, our ideas, our limits by not explicitly asking the simplest and most ontological of questions first and foremost?

Most parents of young children will tell you how the curiosity and wonder of the young means they never stop asking questions. Children call us back to simplicity and help us return to what Caputo (1987) refers to as the "original difficulty." Rather than being consumed with the epistemological—the knowing what and how—children seem most located in asking the ontological Why? Why this? Why that? This one question Why? gets asked repeatedly by the very young. It can be challenging to manage attentiveness to every question. As a teacher of young children, I have lived in that questioning space. I know how it can be. I am not suggesting that as teachers we must respond and answer every question asked by every student immediately and fully. Pragmatically, in a room of thirty children, it simply is not possible.

As teachers we need to know that some questions (mathematical and otherwise) are better than others (Brown & Walter, 2004), but how do we come to know what makes a question worthwhile? How do we come to know our way around the mathematical landscape—indeed the pedagogic landscape—on which we find ourselves traveling with children? In thinking about this, I am pulled back to Gadamer, who sought to explore "understanding" as emerging from "lived experience, Erlebnis" (Smith, 1994). Using Husserl's notion of the life-world, I am reminded of the importance of intentionality of experience. Every experience is an experience of something and not an experience in general. As Smith (1994) states, "Thinking and interpreting are always and everywhere precisely about the world" (p. 108). In this book I explore and unpack my lived experiences teaching children, but I am not the focus of the narratives. Rather, the particular events that I attempt to unpack have something to say about the experiences of teaching, of mathematics, and of schooling in general.

All experiences are not equally educative, as Dewey (1938/1998) states in *Experience and Education*; some experiences are actually "mis-educative." Dewey cautions readers to be careful of "experiences that narrow the field of further experience" (p. 13). Keeping this in mind, an experienced teacher according to Gadamer (1989) will recognize when a question asked opens up a world of possibility. The etymology of the word 'experience' leads to *ex peria*, meaning 'perimeter,' or 'being around.' Thought of in this way, an experienced teacher would mean a teacher who knows his or her way around a topic in such a way that allows the world to become open to interpretation, not to make the world narrower. Experience thought of in this way would not be experiences that narrow the field of further experience.

According to Gadamer (1989), then, being an experienced teacher would mean being a teacher who knows his or her way around a topos. Jardine, Bastock, George, and Martin (2008) explain, "As we become experienced, having cleaved with affection and made ourselves 'roomier,' the world's roominess can be experienced" (p. 53). An experienced teacher, in this sense, is able to see openings for pedagogic possibilities. Clifford, Friesen, and Jardine (2001) describe this form of experience as ecological, as the experienced teacher will "know how to dwell in a place in such a way that things will show themselves in this way, this time around" (¶ 19). Having the same experience over and over again does not make someone more experienced in the Gadamerian sense. The number of times one has the same experience does not open up a world in and of itself. In fact, the belief that 'I've seen and done this all before' can be thought of as narrowing the world, as possibilities for moments of wonder can be missed.

As teachers in our quest to become experienced in a way that opens up the world, we have to pay heed with as much attentiveness to the pedagogic moment as possible, if in that moment rests a child's attempt to ontologically ask the question Why? Paying attention in such a way is also pragmatic. It informs teachers what to do next with the landscape before them. Greene (1995) writes, "All depends upon a breaking free, a leap, and then a question. I would like to claim that this is how learning happens and that the educative task is to create situations in which the young are moved to begin to ask, in all the tones of voice there are, 'Why?'" (p. 6). We cannot hide behind the application of why by asking more questions of 'how' and 'what.' Children learn quickly when their questions are dismissed and will not be answered. Being interrupted by a child with a question that asks, 'Why am I here?' can be an incredible challenge for the beginning teacher. When teachers do not yet know their way around the topic, such as the living discipline of mathematics, they orchestrate learning down a single path and a child's interruption of wondering why can cause plans to go awry. Teachers become afraid of getting lost, do not know where they are on the mathematical landscape, have no way of getting back, and have no way to hear the question. In this instance the child is misunderstood, the question ignored, and life continues for the teacher. In asking why too many times, the only way the situation can be read back to the student is that he or she is the problem. The teacher in this case has yet to realize his or her way around the landscape to recognize there is a place for the question. It is difficult to imagine that it was forty-five years ago when Neil Postman quipped, "Children enter school as question marks and leave as periods" (Postman & Weingartner, 1969, p. 60). I am hopeful that Postman's comment regarding the outcome of traditional schooling will someday become obsolete.

Frozen Future

Greene (1995) suggests that the educative task is to create situations that breed the question Why? Pedagogical approaches that dismiss or fail to entertain Why? questions, present mathematics as certain, fixed, static, and, therefore, not open to the young; indeed, such approaches do not need the young. There is no inheritance, nothing to carry forward. When students ask why they need to know a particular mathematical topic, the answer most often refers to a frozen future, as if time and space were fixed. Teachers unaware of the living unfinished character of mathematics in the world tell their students that they will need to know the topic in question for the next grade or for later in life. It is the promise of 'someday' that never comes. Smith (2006) writes about educating for a frozen future as becoming "more and more of what it always was" (p. 25), and he states:

> Education seems like a preparation for something that never happens because, in the deepest sense, it has already happened, over and over. So built into the anticipations of teaching is a mask of the future that freezes teaching in a futurist orientation such that, in real terms, there is no future because the future already is. (p. 25)

I remember being told as a child the following: "You will need to know your times tables if you want to go to grade four because you will have to do long division then, and you can't do long division if you don't know how to multiply." I was being prepared for a predestined and predetermined future into which I just had to fit. If as teachers we want to educate children for the future, then we must realize that the future is not frozen and neither is mathematics. Being told to do something so that I could use it in the next grade was not good enough thirty years ago, and it is not good enough today.

What is troubling is that as teachers we often forget, ourselves, to ask why. Why am I doing this investigation? Why is this topic important? Why is this mathematics worthy of anyone's attention? Why am I doing it this way? Why am I here right now in this place with these children? I am continuing to examine the question 'Why?' more fully of myself. I am attempting to figure out why I believe what I do about children, about mathematics, and about schooling. In my process of coming to understand, I am working to "fuse my horizons" (Gadamer, 1989, p. 307) as I try to weave among the tensions of the strange and the familiar, the established and the new. In working to make sense of these topics I am working toward a deepening understanding.

Pedagogic opportunities occur over and over in classrooms but to the beginning teacher many wonderful moments, ripe with openings to new understandings, are not noticed or are overlooked. As we gain experience in the Gadamerian sense, we can attune ourselves differently to mathematics. We gain the ability to see the openings, and, it is hoped, to have the courage and the imagination to be attentive to them. If the moments are recognized, something interesting may happen that leads learners and teachers somewhere, and if the moments are not attended to, then the child may 'learn' lessons about dismissal, injustice, positionality, and loss. Maxine Greene (1995) speaks of "openings to possibility" (p. 5) through listening and attending to children. This book is my way of recovering for the reader and for myself what has led me in my journey to understand teaching children as well as what has brought me back to the original difficulty (Caputo, 1987) of my teaching practices.

In the next two chapters I wish to point to particular moments of pedagogic awareness that I hope will crack open for the reader the living nature of teaching children and, in particular, teaching the living discipline of mathematics. I am, as Jardine (2006) describes, "working such matters out" (p. 161) through my writing to come to a deeper understanding. In chapter 6 I describe experiences from a classroom of grade-one learners I once taught, and, in retrospect, they have taught me more than I did them, and chapter 7 focuses on learner teacher classroom experiences in grade two.

IF 5 IS THE ANSWER, WHAT MIGHT
THE QUESTION BE?

Ms. Stordy, I want to come talk to you about math and grade one. Can we arrange a time? Thanks, Susan. There it was, a note written in William's agenda in ink beside Tuesday's agenda message. In the busy pace of the school day, the scanning of agendas for messages from parents becomes another teacher reconnaissance skill. While keeping an eye on all children and ensuring they are 'on task,' the skilled teacher is able to get to every student's agenda to check for notes from parents about hot dog money, early pickups, missing jackets, and, now and then, curricular issues. I was not one of those skilled teachers described above who could organize her day in such a way that the checking of agendas was made without some effort. It seemed to be one of my least favorite activities because it cut short learner teacher work and play and the surveillance task—checking journals—was, therefore, often left to my last minute. While reading Susan's parental note I was standing in the midst of a dozen other agendas being thrust at me—I felt as though all the spokes led to me, the center/hub of a wheel. The children were bustling about getting their coats and boots on to catch the bus for home. 'What does Susan want to know?' I thought. 'We're following the *Program of Studies*. It's grade one, for heaven's sake. Her son William seems to be a good mathematical thinker, at least from what I can tell so far. What does she want? Now I have one more thing to deal with tomorrow in a day that is already packed.' I juggled the agenda onto a desktop free of backpacks and scribbled underneath her note, my left hand now smudged with the ink from the pen, "Hi Susan. Can you come by while the students are in gym? Say, 11:05 a.m.? We can talk then!—Mary." William waited impatiently, dancing from one foot to the other, bursting out of his winter coat zipped tight. "I'm gonna miss the bus, Ms. S!"

"Plenty of time, William. Here. Be sure to give your mom your agenda when you get home."

"Yup! Thanks!" he replied, and together we stuffed the agenda into a backpack crammed with library books about armor and knights. Within a flurry of minutes, the room was silent with only the far-off sounds of the diesel buses growing more distant. Another day with the children was over.

As for most teachers, for me, with the quiet of the empty room comes a time to replay the day and to bring to the forefront the issues requiring the next moment's attention. After robotically picking up scraps of paper off the floor and old notes from the post office from under desks and every corner of the room, I could feel a tension in my body. I was not at ease. My mind came back to the issue that was causing my restlessness. Right. The note. William's mom. Math. Up to this point we had being doing grade-one mathematics in a variety of ways, in ways that seemed natural to me and that certainly covered the curriculum outcomes. What might Susan's questions be? What expectations does she have that are not being met? How do I prepare for this meeting?

ACT 1.   Scene 1. Mathematics.
            [Adult teacher and twenty-five grade-one children stand in a group in the center of a crowded classroom. Children are moving. Some are bouncing on their toes. A few crouch down and tie shoelaces. Most are excitedly whispering to one another and negotiating a spot from which they can see the teacher and one another. The whispering in the room drifts to a silence. The children stand almost completely still.]

TEACHER.   So, if the answer is 5, what might the question be?
            [Hands spring up from most, if not all, children. Stifled squeals erupt. Teacher nods at Joseph.]

JOSEPH.   What do you get when you add 3 and 2?
            [Teacher waits for challenges to this question. Heads bob up and down in agreement.]

CLASS.   Yep.
MAXINE.   That question makes sense.
TEACHER.   Another question? [Gestures to Brianne.]

BRIANNE.   What do you get when you add 4 and 1?

CARMEN.   What do you get when you add 5 and 0?

DARCY.   What do you get when you subtract 2 from 7?

WILLIAM.   What do you get when you add 1 to 8 and subtract 4?

              [*Fewer heads bob right away, as children mentally do the calculation. Teacher sees this.*]

TEACHER.   William, it would be a good idea for you to explain your question to the class. Can you show us this question and convince us that your question makes sense? [*Teacher gestures to Maxine.*] You work with William to show us this question, will you?

              [*William and Maxine whisper quietly as they appear to make a decision on how to proceed with the problem. William corrals nine children away from the main group with his arms open wide. The nine children comply and move away from the group.*]

TEACHER.   Ready? [*Maxine and William nod.*]

MAXINE.   If we have eight people … [*William takes eight students one by one to the other side of the room. As he does, the larger group of children counts out each student as one big chorus.*]

CLASS.   One! Two! Three! Four! Five! Six! Seven! Eight!

MAXINE.   And we add one more … [*William takes the ninth person by the hand and walks her over to the group of eight.*] We have nine people. [*Maxine and William now trade positions quickly.*]

WILLIAM.   So, you can see we have nine people now. 8 + 1 = 9. [*He turns to the whiteboard and with a blue marker puts the equation on the board.*] Does everyone agree?

CLASS.   Yup!

WILLIAM.   Okay. But if Maxine takes four people away [*Maxine taps four children on the shoulder and ushers them to the other side of the room. As she taps them, the class counts out loud One, Two, Three, Four!*] that means we have five people left. And as you can see, that is how many people are over there. [*He points to the group of five. Maxine is now back at the group counting one by one by patting on the head.*] One, Two, Three, Four, Five!

TEACHER.   William, you wrote on the board 8 + 1 = 9. What would your equation look like now that you removed four people?

              [*William scratches his head, and then writes on the board: 8+ 1 − 4 = 5*]

WILLIAM.   This is what we did. We took eight people [*points at 8*] and added another person [*points at 1*] and then we removed four people [*points at 4*] and when we did all this what we were left with over there was five people. [*He points at the symbol 5 and the five people left on the other side of the classroom.*]

TEACHER.   Thank you for the demonstration, Maxine and William. Thank you to the participants. [*All children come back together as one big group.*] Can anyone think of another way to make five that we haven't talked about yet?

Learning with and from the Parents

"I don't see any work sheets coming home, Ms. Stordy. My neighbor's son is in the same grade as William and he brings home his work sheets filled with arithmetic for homework, like my older son used to get, and like I used to do. I asked William about math and he says you don't do it! Where are William's worksheets? Surely you are doing arithmetic with the children! So why is he not bringing his sheets home?"

In the crowded classroom we sat knee to knee on a pair of chairs too tiny for adult comfort. Cardboard boxes of various sizes that were in transition to becoming castle walls sat all around us. When gym time arrived the children had simply suspended their work where it was. They had just been delivered to the gymnasium at the other end of the school. I had twenty minutes left of my prep time to talk with Susan before they would be back.

As I sat listening to William's mother's questions, I remained quiet but inside I was anything but quiet. Inside I was playing out various ways to answer her. Where were the worksheets?

To answer her, I had options. As a relatively new teacher who looked younger than my years, I could turn to research on teaching and learning. I could turn to scholars such as Dewey, Ball, Towers, Greene, Boaler, Mason, Jardine, Davis, and Friesen. I could talk about inquiry and collaborative learning. I could choose to default to becoming the voice of authority. I could hold on tight to what Britzman (1991) refers to as the cultural myth of the "teacher as expert." I could talk about how mathematics is more than symbols on paper to override this parent's possible accusation that maybe I didn't really know what I was doing. I was the teacher. I was the professional. Why was I being questioned? Then I remembered an excerpt from the book I was given by my administration upon my arrival at the school earlier in the year. P. Palmer (1998), in *Courage to Teach*, writes about authority, and he states:

> Authority is granted to people who are perceived as authoring their own words, their own actions, their own lives, rather than playing a scripted role at great removal from their own hearts. When teachers depend on the coercive powers of law or technique, they have no authority at all. (p. 33)

By using my authority of teacher as a position of power, I would be playing a scripted role. It did not feel right. I had another option. I needed to author my own actions. In this case, I needed to corral my ego, stop planning my defense, and try to understand the parent's concerns; indeed, I needed to consider her Why? question.

ACT 1.	Scene 2. Mathematics Continued.
	[As the scene opens the room is buzzing with chatter. Children are up and moving about the room, working in small groups. Two children are at the board showing something to two others. The teacher is sitting on the floor near the post office and coat hook area with a group of four children who are squatting on their legs like frogs, working with balls of play-dough spread out in some sort of grid on the noncarpeted section of the room. Teacher looks at the clock. Snack time.]
TEACHER.	Okay, everyone! I want you to just leave everything as it is right now. Pretend you've taken a picture of the moment, and we will return to this moment after recess. Right now, I would like you to get a healthy snack out of your lunch kits.
ALICE.	Is it really snack time already? It must be, because I'm starving!
TEACHER.	Yes, Alice, it most definitely is. You've been working pretty hard so we'll take a break now. Please take your snack to your desk or a friend's desk. Careful where you walk, remember! We'll continue working on our math after you come back inside.
	[Fade to black.]
	"Susan, why don't you come hang out with us Monday after recess? I think when you see what we do you might have a better sense of things as William experiences them. I would love it if you could come join us. I know William will be pleased. What do you think?"

The Seven-ness of Things

William's mother did join our grade-one class on Monday. When we came together as a whole group, she sat among the children and listened, cross-legged like the rest of us around our legless, unfinished round table that sat on the floor. Another parent was also in the group as the scheduled classroom helper for the day. Since my teaching practice did not revolve around work-sheets, my parent helpers were rarely given the task of being sent off to the copy room to stand next to a photocopier. Instead, I wanted them not only to be engaged in the work of the children, but also to contribute something useful for me as the teacher. One of the tasks I would often assign was that of scribe; that is to say, the parent would write down verbatim what was said by an individual, a group, or the whole class without interpreting or changing the language. With younger children who are at the beginning process of writing it is not easy to capture the depth of their thoughts and ideas if left to their own developing writing abilities. As a group we came to new understandings continuously through conversation, so often I would ask parents/guardians to scribe our conversations. They would help turn the oral text into printed text. Isabel's father was today's helper, and he sat with marker and paper in hand. Having this task completed allowed me to focus on what was being said, who was saying it, and to be more responsive to the children without trying to hurry to get down their words on paper. I would often place the written text of our conversations on the wall for future reference and for the children to work with as another printed text to read. In this case, it was their collectively created text about mathematics and the 'seven-ness of things' that would soon be taped to the wall.

What follows is a peek at the type of scribed conversation that helpers would capture for me on chart paper. The speakers are only identified as either teacher or student. 'Student' represents many different speakers.

TEACHER.   What day is today?

STUDENT.   Monday!

STUDENT.   Yay! It's Moon day! Are we going to read another story about the moon today, Ms. S?

TEACHER.   That's right, Brent! Monday was named for the moon. You remembered! Perhaps you can go to the library when we have some free time and ask Ms. Lilo to help you find a good moon story for after lunch. But right now I want to check to see how everyone knows it is Monday. Wasn't it just Monday a few days ago? Are you sure it's Monday?

STUDENT.   Well, yesterday was Sunday…

STUDENT.   Named after the sun!

STUDENT.   Tomorrow is Tuesday so today has to be Monday.

STUDENT.   Can I bring in a story about Mars tomorrow for Tuesday, Ms. Stordy?

TEACHER.   That sounds like a great idea, Hannah.

STUDENT.   The calendar's right there, Ms. S. you can see that today is Monday, the twenty-first.

TEACHER.   Oh. It was Monday just a few days ago wasn't it?

STUDENT.   No. Not a few days ago.

TEACHER.   Well, when was it Monday?

STUDENT.   It was Monday seven days ago, Ms. S. you know that. We all know that!

TEACHER.   Yes, you're right, Isabel. I do know that. I was being silly, but it did feel like it was only a few days ago. Can you explain how you know last Monday was seven days ago? I didn't see you counting anything.

STUDENT.   Because there are seven days in the week. And when I go back using the calendar… 20, 19, 18, 17, 16, 15, 14, I land on Monday the fourteenth.

STUDENT.   And it will be Monday in seven more days, Ms. S.
STUDENT.   And in seven more days after that.
STUDENT.   And after that. And after that and after that and after that…
TEACHER.   Okay. I get the point. Mondays come every seven days. That is what
            you are saying. But what about Tuesdays?
STUDENT.   Every seven days. You are being kind of silly. We already know this
            stuff.
TEACHER.   Yes, you do. These questions are way too easy. I guess I was wondering
            then if you were able to tell me other things you know about seven.
STUDENT.   You mean like what makes seven?
TEACHER.   I mean whatever you can think of about the number 7.
STUDENT.   Um… 6 and 1 make 7.
STUDENT.   5 and 2 make 7.
STUDENT.   3 and 4 make 7.
STUDENT.   4 and 3 make 7.
STUDENT.   I said that already.
TEACHER.   Is that the same thing?
STUDENT.   The answer is the same, and no matter which number you say first it
            still makes 7.
TEACHER.   How do you know?
STUDENT.   I think we should include it anyway. We should write down 1 and
            6 and 2 and 5 then, cause we said them already. See? Isabel's dad has
            'em there!

By providing you with a brief example of our dialogue above, readers may understand that we had been discussing a number, in this case 7, but they may not necessarily notice the important mathematical concepts that had been taken up by the children. The conversation above led into the realm of commutative properties. On this day the children discussed whether or not they were wasting time by writing the same two addends down; they noticed that it did not matter which number came first. Following the excerpt above, the conversation continued. Multiple numbers got added to make 7, which focused the discussion more on order property. It was suggested that $2 + 3 + 2 = 7$. The entire class agreed that it did not matter whether the 2 or the 3 or the other 2 went first or second or third since they all add up to 7. Soon the chart paper was filled with equations in which 7 was involved. To my delight the children began to create equations such as: $x + 2 = 9$ and $10 - x = 3$ to show 7 as the x. There was a moment of silence while the idea of whether or not the order of the numbers matters when working with equations involving subtraction. Someone asked about $9 - 2 = 7$ and whether or not you could write $2 - 9 = 7$. The majority of the group said it was impossible, except a few who claimed it could be done. From this example, I knew I had work left to do with everyone in the room in deepening understanding around the very difficult concept of subtraction that so often gets simplified in classrooms to 'no, you can't do it' or the more frequent untruth, 'the bigger number goes on top.'

TEACHER.   I have another question for you. How many dwarves lived with Snow White?

STUDENT.   Seven!

TEACHER.   So, I want you to think for a minute about what else you know about seven.

STUDENT.   I have two brothers and one sister and my mom and my grandma and my dog, Buddy, so that makes seven in my family.

STUDENT.   Hey! Ms. S there is seven letters in the word 'Tuesday'!

STUDENT.   And seven letters in my name! Michael!

STUDENT.   Seven is the number on my dad's hockey uniform!

STUDENT.   There are seven days of the week of course, Ms. S!

The conversation carried on for a long time as the children's experiences with 'seven' began to pour from their lives. I was told about neglected fish in tanks at home and spots on ladybugs from a book that sat on a desk. We counted seven holes in our head and talked about years of bad luck. We looked at dice and saw that the numbers on the opposite sides of the cube always add up to seven. We got up and walked around the room while keeping control of our bodies and kept from touching anyone, and got into groups of people that represented our answer as quickly as possible when I shouted out a math problem. For example, I shouted out "What is five plus two?" They quickly and without using voices (other than the little squeals that usually accompany such activities) got in groups of seven. I had the two parents join us to make an even twenty-eight. What surprised me was how the children held the concept of one week being equivalent to seven days. We came back to the table and turned to the calendar once more. Using the calendar the children stopped counting days and switched to weeks as we talked about how many days there were from one date to another. The concept of one becoming seven is not an easy concept, but within minutes they were pointing out on the calendar how two weeks and two days was the same as sixteen days. This far exceeded my expectations for what I thought we would cover regarding the curriculum standard for number relationships. William and his mother exited the room on their way to a local restaurant for lunch; she turned back and met my eyes.

"Thank you!" she mouthed and smiled.

"We'll talk soon," I mouthed back over the sounds of the children as they compared sandwiches and treats.

The Conversation Must Continue

While initially appearing disruptive to my teaching, Susan's questions about the mathematics that was actually happening in our classroom turned out to be one of the best gifts I could have been given as a classroom teacher. It forced me to look at what I had been doing with mathematics in a much more critical way. Up to this point, a third of the way through the school year, I did what I did intuitively, as it felt right to cover the curriculum this way. That is to say, it felt right in an embodied phenomenological sense to expose children to the idea that mathematics is about more than one right answer. It felt right to explore number relationships in ways other than work sheets that simply showed how and what.

My experience with William's mother showed me that I had to do a better job of providing evidence of understanding and I needed to think about how to help parents begin to reconsider what constitutes evidence of learning. It asks a lot of parents to support a move from something they understood, such as worksheets, to something quite unfamiliar (Kohn, 1999). Completed pages of arithmetic seemed to be an unquestioned source of solid evidence of understanding. After witnessing William engaged in number relationships in a variety of ways and Susan's gratitude for being allowed entry into our space of mathematics, Susan seemed to recognize some of what my intentions were; but I knew the conversation had to continue.

I began to have conversations with the children about mathematics, because it became clear to me after my experience with William's mother that how I was doing math was challenging to parents and to their image of what mathematics might be. It was at this time that I recognized that I needed to be clearer about how I was meeting the curriculum outcomes for grade one through the types of activities we were undertaking, and I needed to extend my work to families so they would understand what it was that we were trying to do. My journey of making sense of mathematics and children was still really beginning and now it was branching out to the political realm for the first time in my teaching career. It became clear to me that I was not only working with a whole group of children to help them engage with the living character of mathematics and to help shape what mathematics was for them, but I was also left with the more challenging task of trying to help parents understand.

· 7 ·

THINGS ARE NOT ALWAYS AS THEY SEEM: ENTERING THE CLASSROOM

The work of the children and teachers at the public school in rural western Canada where I was teaching meant that my work as a teacher was often in the public sphere, as was the work of the children. When I began teaching grade one, I did not really understand how public it would become, but such a stage forced me to re-think (a) what I was doing on a regular basis and (b) why I was doing it; almost certainly someone would ask these questions. Prior to teaching at the school, the only time I had another adult in my classroom was when a principal was evaluating my teaching. However, from day one at this western Canadian school, doors were open to parents/guardians and to any of the many visitors to the school from other parts of the province, the country, and the world. Guests made their way from as far away as Africa, Asia, Europe, the United States, and Australia. Visitors were educators, policy makers, and politicians as well as people interested in technology, educational reform, innovation, and educational research. Many visitors were parents of school-age children who were planning on moving to the area; in some cases families were moving so their children could attend the school, as the profile of the institution had become more recognized. The work of the children and teachers began to appear in the media, and my time teaching at the school afforded me the opportunity to talk about teaching practice with national and local newspapers as well as mainstream news magazines and international researchers.

I want to invite readers to visit the time and space of our grade-two classroom and to experience the complexities of our learning. What follows is a narrative simply called "A Foreign Land." It is structured in such a way that disruptions—breaks to other texts—are placed within the main narrative. This requires readers to go back and forth between texts as they journey through a visit to the classroom. I hope to bring forth elements of our work to readers that will help uncover what the students and I were doing, and, speaking to the broader truth, what children are capable of doing when allowed to be in school and with mathematics in this way.

A Foreign Land

The visitors entered the school for the first time and stepped into the Gathering Area.[6] On this day, like most, the large open space that greets all people to the school was full of natural sunlight streaming in through substantial windows. The windows served as giant picture frames for the towering lodgepole pine and aspen trees in the wooded space behind the school. Impressive log poles supported the high open ceiling in the same architectural fashion as other parts of the building constructed in the mid-nineties. The visitors noticed a group of children, about age ten, working on the floor of the Gathering Area. The children appeared not at all fazed by the guests' sudden appearance, and after giving a quick smile and hello, they continued on with their business at hand. Within seconds of entering the building, a smartly dressed woman with a kind smile issued a warm welcome to the visitors. As the principal of the school, Ms. Rose gave a brief history of the place and shared the usual information that many visitors wished to know, such as the intentionality of the architecture of the school to serve the pedagogic needs of the students and teachers. She explained the four pillars of the school upon which most of our work was based: inquiry-based learning, classroom-based research, integration of technology, and staff professional development. These were the elements that guided the goals of the school. There had been visionary educators employed at the school when it first opened, and because of the work of those pioneer teachers and administrators, much national and international attention was now focused on the school. It was important for visitors to know that the place, while not yet five years old at the time, already held an important history.

Soon, the guests were escorted a short distance through a wide-open hallway area containing clusters of computers. There were children of various ages working at desktops. There was a quiet buzz in the hall as children discussed their work. Similar to the children in the open space at the entrance to the school, these children did not seem concerned by the presence of visitors. Rather, they seemed eager to explain their work, to respond to questioning looks, to explain what they were doing. A young girl with a long, dark-brown ponytail was down on all fours reaching to the back of the computer. She was holding on to a headset with a microphone. "I want to put sound into my story," she explained to the visitors, although they had not asked a question of her. She smiled a broad smile that revealed a noticeable gap in her front teeth. She joined her partner, and the two children quickly immersed themselves in their task at hand.

Beyond the cluster of computers in this hallway was a room that did not look like an ordinary classroom as popular culture might depict in schools. There was no obvious front to this room. At first glance, there were no desks in rows or anywhere else, for that matter. There was no teacher's desk and no chalkboard. The room was full of activity but there was no obvious tall adult leading the group in unified action. In fact, at first the visitors wondered whether this was indeed a classroom at all, since there were large papier-mâché objects in the room and dark fabric covering the ceiling. A tiny boy in a bright red T-shirt holding a Lego wheel in one hand bounced up from a cluster of students working on the floor to meet the visitors at the doorway.

"Hi!" he said, "Welcome to our classroom! Come on in!" The visitors smiled at the boy and stepped through the open doorway. It was not an ordinary-shaped room; the back wall was wider than the front wall. A large window was set into the shorter inside wall by the doorway, allowing for an unobstructed view of the working area in the hallway from the classroom. Visibility was an important component of the school, Ms. Rose had told them in their brief introduction. It blurred the boundaries of learning spaces and allowed children and teachers to work in various locations.

Shifting Locations: Shifting Perspectives

Often the children were spread out between the various spaces, doing many different things depending on the topic of wonder. It was as if the classroom had no real boundaries. Boundaries to the outdoors were blurred as well, as forty acres of a forest surrounded the building as part of school property. The children, with my teaching partner, Cheryl, or me, would spend time exploring the forest, looking at animal tracks, and becoming observers of patterns. Together we noted the passing of time through the mark of the seasons, the life and dormancy of plants, and the recurring patterns of the moon. We looked at natural shapes and structures in the life of the forest. We observed the differing patterns of ice that coated small puddles and the top of the pond that sat between a grove of trees.

Sometimes I would sneak the children up on the roof so we could use the flat surface of the school as part of their learning space. It was a dry place to sit and observe the sky and the land, unobstructed by lodgepole pines and towering aspens. It was also a place to sit and to write. Perhaps most important was that it helped shift our perspective of what lay in our view from up high. One of the things I tried to do with the students was to agitate, wrestle with, and create ways to have their perspective shift from themselves as the center of their universe to a new center temporarily. Back in the classroom, I wished for them to consider the world from another point of view. For this reason, the sky filled with constellations on a dark blue sheet of fabric on the ceiling was the sky as seen as if we were on Mars. It was planet Earth painted blue that dangled by a piece of thread from the ceiling. The moons, also suspended in the Martian sky, were the moons of Mars, not Earth's moon. They were not smooth and round, as Earth's moon would be, but rough and ill formed in papier mâché, to be as true to the odd shapes of the two moons of Mars, which lack enough mass for gravity to assert its pull toward their centers. I never set out to create the moons of Mars. In fact, I did not even realize they existed, despite having done my own research on the planet. Rather, it was the students who discovered the moons in their research and insisted it was important to include in our skyscape. They even found out the names of the moons, Deimos and Pathos, the Latin derivatives for 'fear' and 'panic.' We discussed why these names seemed appropriate for exploring unknown territories. Seeing those moons dangling under the Martian sky reminded me to challenge myself with my own perspectives.

Teaching as a living discipline can feel like exploring unknown territories. In some way those moons represented for me what I now know to be Gadamer's (1989) notion of believing there might be truth in what the other has to say, regardless of how foreign such a perspective or viewpoint might be from my own. At the time of teaching grade two, I had not yet been introduced to Gadamer's work. I was still learning [am learning still] a great deal about teaching children. I was not able to articulate why I wished to challenge the students and myself with reminders that different perspectives need to be considered and challenged. Perhaps it began when my own assumption about teaching spaces was unearthed. This was brought to light for me by witnessing other students and teachers at the school the previous year. Once I was able to recognize that our learning space could be any space that made sense, such a simple realization opened up for me a new way to think about learning environments and the impact that the space of learning had on what could be learned. The physical location of my students and our work was occasionally a topic of discussion with visitors. Unfortunately, such discussions often came around to issues of liability and rules rather than the richness that other locations, such as the forest, afforded us. This was disappointing, since the pedagogic reasons were not seen, or not seen to be powerful enough to override issues of control and conformity. For these visitors, the place they were seeing was not the place of grade two with which they were familiar. Rather, our place of grade two was a foreign land.

(Back to A Foreign Land)
The wall to the right of the entrance was retractable. On this day it was open one-fifth of the way, exposing an expansive room beyond it with a sink, large tables, and many more clusters of computers. Back inside the classroom, the retractable wall was covered in various printed texts on different pieces and sizes of chart paper. One piece of chart paper had a problem on it that read:

Dear Grade Two,

If you took one step backward for every two steps forward, how many steps would you need to take to go from one end of the Gathering Area to the other? Let's figure it out!

Ms. S

The Gathering Area: Two Steps Forward and One Step Back

To the outsider, I suppose it looked more like line dancing or gym class than mathematics. We were in the Gathering Area of the school and were spread from one end of the wide-open space to the other. The well-known expression about taking one step forward and two steps back had gotten us thinking. I was trying to help the children understand numbers. I wanted to begin to shed the notion that numbers go in only one direction, beginning at zero. I also wanted our/their learning to be based on the idea that good questions require thought, that mathematics is not about getting the right answer immediately, and that a good question does not always have to be applicable to the real world but it should certainly allow learners to think mathematically. In other words, learning to think mathematically was important to foster in everything we were doing. I wanted to have the children do mathematics wherein addition and subtraction were deeply connected, and wherein moving our bodies as a way to solve the problem would be essential to helping us think through the problem. In this case of stepping forward and backward, I had placed the problem on the wall in the classroom early in the morning, and as they arrived the children were to be thinking about what exactly the question was asking of them. If for every two steps forward, you took one step back, how many steps would you have to take to go from one end of the Gathering Area to the other?

Once the entire class had arrived, we decided it was necessary to go to the Gathering Area to do some figuring out of the problem. This question was simply the jumping-off point for thinking about other possibilities and combinations of steps as we worked together. The children decided that a step forward was equal in size to a step backward. We negotiated whether my answer would be the same as theirs since I was taller and took bigger steps. We made up other problems, some of which seemed impossible for advancement (one step forward and one step backward) and some children even changed single steps into halves. After we had explored many possibilities, we returned to our classroom and the children captured in their math journals their own discoveries about steps and numbers in this problem. While I did not explicitly talk about a step forward being +1 and a step backward being −1, I did hope it planted a seed concerning how numbers and their relationships might exist in the world.

(Back to A Foreign Land)

Positioned by the retractable wall were what appeared to be the remnants of castle walls similar to the ruins one might find when exploring the countryside of the United Kingdom. Unlike UK castle ruins, these grayish-black walls were void of stone and stood about five feet high. Recycled cardboard boxes from the families of the children who had been in grade one together contributed significantly to the longevity of these particular papier-mâché castle ruins.

Castle Remnants: Keeping the Conversation Going

Staying with a group of students for two years, referred to in schools as 'looping,' was strongly encouraged by the school administration, particularly when engaging in inquiry-based teaching and learning and curriculum. As mentioned in chapter 1, I was able to teach grade two to many of my grade-one students as well as twelve new students, thus beginning the teaching year with thirty children. It was a wonderful way for me to continue our work from the previous year, and to continue working in the way that we had been. With twelve new students it meant bringing them into our way of doing school. There were adjustments; students used to other classroom routines and children from outside the province had joined our class. Most of the adjustments for the children dealt with them re-learning what school might actually be. For one out-of-province child, Jordan, who previously had math worksheets every day at school and then most nights for homework, I heard him declare to his mom, who was our helper one day, that "we don't do any math" as he programmed a robot to stop automatically through the use of a touch sensor.

At the age of seven he arrived in my classroom and appeared to have a strong opinion about what school was. From my conversations with him and his parents, doing school seemed to be doing worksheets, sitting still, working individually, printing neatly, and passively doing what was asked of him. Worksheets not completed in class meant worksheets were done at home, which meant he and his mom struggled every night to complete his math work. It took time for Jordan to realize that work could be enjoyable, and that siloed subjects might not be readily discernible when working with topics such as the work we completed in grade two on the exploration of Mars. Our day was not about math, science, language arts, or social studies, but simply our work was about the exploration of Mars. As the teacher it was my professional obligation to see that each discipline in the *Program of Studies* was being addressed, and as a lesson discovered from grade one and William's mother, I realized it was important for me to explain to parents through regular curriculum letters sent home exactly how the school subjects were being addressed in our inquiries.

The remnants of the castle from our investigation into life in the Middle Ages in grade one stood as a reminder to the children about our previous year's work, and also of how we worked. Housed in what might appear to outsiders as a heap of cardboard, glue, and paint, was the memory of our work together, but it was much more than that. We were cultivating our collective memory. Jardine, Bastock, George, and Martin (2008) write:

> The cultivation of memory is a way of working in the classroom, a way of caring for the living worlds of knowledge we are inheriting and caring, too, for teachers and students alike in their working their ways through those worlds and the nature and worth of those inheritances. (p. 33)

The cardboard and paper creation represented hard work and vision. It stood for the idea of being able to stick with something when it might have been easier to forgo and create a less complicated, less messy structure. Sitting in the grade-two classroom, the castle walls served as our backdrop but were also a constant reminder. We took a piece of grade one with us. Memory is a way of handing things along. Our castle walls were a kind of family artifact that was being handed along. We were a different class because of the work we did together in digging into medieval life. The memory was housed in us but also in the castle walls that served as our totem for what was possible.

(Back to A Foreign Land)

Exploring the Martian Landscape: Creating New Terrain

As previously mentioned, communicating with parents and guardians was something that I realized was critical when working in nontraditional ways with children. Ever since my conversation with the parent Susan in grade one about whether I was doing any mathematics, it became clear that I needed to be proactive and explain what I was doing and why I was doing it. It was important to send letters home regularly and to post them on our class web site. The letters often explained upcoming events and how the work we were doing in the class was related to the curriculum outcomes as stated in the *Program of Studies*. What follows is an example of a letter in which the purpose of the Martian terrain is explained:

Dear Parents/Guardians,

Early in the school year, we spent a lot of time focused on humanity's quest for exploring unknown lands and worlds. To begin examining the drive for conquering new territory, we looked to the past and read one of the oldest stories to date, Homer's The Odyssey. *By following the adventures of Odysseus as he tried to find his way home, we spent a great deal of time learning about how we tend to fear things that we do not understand. We read many of the stories of the ancient Greeks. The mythology of that time was a way of humans explaining their world and making sense of the world. This work was connected to our round table and to the skyscape; the children learned about the Greek myths attached to the constellations of the zodiac. They represented them on the table and on the fabric that covers our classroom ceiling. As we took up humanity's drive to explore, we looked at modern historical exploration and made many direct connections to European colonization of North America. Issues connected to the environment such as clean drinking water, natural resources, and pollution were topics we wondered about together. We then cast our eyes to the present and the future with regard to exploration. As you know, NASA has a team of many scientists researching the exploration of our neighboring planet, Mars. My students, your children, may very well live to see the day when travel to Mars and the colonization of Mars could become a very real possibility. Thus, as a class we are researching how we might begin to explore this planet. Like NASA, we are designing robots for specific purposes. The children have broken themselves into teams of researchers who are investigating very real questions. Please see our class web site for a complete list of the beginning questions from which each group is starting to work. The search for life group has questions around whether or not a seed is living. The search for water group is wondering about the presence of ice on the polar ends of the planet. The group that is designing a robot that will dig underground is working with the assumption that mysteries remain below the surface of the red planet. The analysis of air group wonders what kinds of plants and life might flourish in different atmospheric conditions. Please ask your son or daughter to talk to you about his or her group's work. As you know, you are welcome to join us any time in the classroom. Your child may also take home his or her journal depicting progress through the work.*

Some of you were wondering why I asked for donations of plywood and newspaper. Our class has joined a project supported by the Jet Propulsion Laboratory (JPL) in California called Red Rover Red Rover.[7] *We received funds from the school to be able to participate. This particular project allows us to work with schools all over the world, which are also a part of the project. Every class is asked to create a simulation of the Martian terrain and then, through the Internet, special software, and a web camera, we can manipulate robots remotely on the terrains of other schools in other parts of the world, much like NASA's scientists will work.*

We have a map of the planet Mars and have spent time understanding the various areas of the planet. The children have chosen to create the section of Mars that contains Olympus Mons, the largest volcano in the solar system. Through the help of Dr. Karen Dempster, we received advice from a mathematician at the university who determined the appropriate scale for our section of the terrain. The terrain is currently being constructed by the children. We have been working with the concept of scale for a while now. They seem to understand it a little more when working with digital images on the computer. They know that decreasing their image 50% means making it half its size and that all things 'shrink' equally in proportion to one another. The height of the mountain must be proportional to the heights of some of the other hills and valleys that we are creating. We began the work by drawing a 5 × 5 grid on our frame and we have been working on each square in the grid to make sure it is to scale to the identical grid on our map of Mars. If you check your child's math journal you might have noticed attempts to draw items on our desks to scale. If you did not see the math journals you might recall the agenda message sent home last week about standing on top of all of our desks and looking down. The movie Honey, I Shrunk the Kids *actually confused some of the children with their beginning understanding about scale, since everything in that movie does not shrink proportionally. It has been interesting to see the connections the children are making between fractions and percentages and this understanding has come from their work with digital images. For example, they seem to understand that 50% is the same as half. They also find it interesting that 25% is the same as a 'quarter' and that 25 cents is also called a quarter! I am trying to have them see that the language around them comes from somewhere, especially language used in math.*

You will find on our class web site a direct mapping of the curriculum outcomes to our Martian exploration project for language arts, science, mathematics, social studies, fine arts, and health. If you have any questions or comments, please come see me. You are welcome at any time to be a part of this work as the children continue their investigation. Thank you for your continued support of your child's learning.

Until next time,
Ms. S

(Back to A Foreign Land)

On the terrain rested half a dozen Lego robots, none of which was exactly alike. Perched on top of one robot was a portable web camera. On a computer monitor beyond the terrain the visitors suddenly noticed that their own image was being displayed as they passed in front of the web camera. Similar to other visitors to the class, these visitors appeared surprised to realize that their own image was now part of the immediate life of the classroom. They, who had come to see the class, were also being seen, and in this case from more than one vantage point. Who was watching whom?

One tall visitor conscientiously brushed her bangs and tucked a stray strand of hair behind her right ear. She quickly stepped outside of the camera's range, and felt something hit her head. She looked up. Suspended from the ceiling above the mass of red glue and paper was a dark sheet of fabric. The rest of the visitors stepped farther into the room to join their colleague and looked up to see what she had noticed. There on the midnight blue fabric were painted constellations in silver paint. A blue ball resembling a far-away planet hung from the skyscape, as were two oddly shaped gray balls. It was at this point that the visitors saw that there were indeed desks in the classroom but they had been shoved to the perimeter of the room. A sheet of paper with the numbers 1 to 100 listed in ten rows of ten (commonly referred to as a Hundred Chart) was taped to the corner of every student's desk; a larger identical chart was taped to the wall to the left of the doorway.

Beyond the red terrain was a cluster of computers; one was displaying the view from the web camera on the robot on the terrain and one was attached to an LCD projector and electronic whiteboard.

Behind the computers in the very back corner of the classroom was what must have been the teacher's desk as it was coupled with the only adult-sized chair in the room. The desk was covered with books and paper and resting on top of the pile was an open notebook filled with handwriting. A colorful Hundred Chart poked out of a book with a Greek-sounding title, *Eratosthenes*, and it sat next to the notebook.

The Notebook: Working Alongside Eratosthenes

If you were able to pick up and read the notebook you would find a lesson plan, and in this particular instance you would find notes about the Sieve of Eratosthenes. The importance of this lesson is that it demonstrates that mathematics comes from somewhere, that it has a human story, and it allows what some might consider difficult mathematics to be brought to the classroom and made real for children. Through this lesson the children did what the mathematician Eratosthenes did. What follows is an example of my written plan to myself, including a Hundred Chart and a brief reflection on the lesson:

> *Sieve of Eratosthenes*—I need to let children in on this interesting mathematics as well as become exposed to some of the mathematicians who lived many years ago. By using the following activity I can let Eratosthenes enter the room! I know that identifying the first one hundred positive prime numbers is not a specific objective in the grade two curriculum, but this particular activity is all about numbers. By working through this activity the children can continue to work on number relationships and to recognize patterns, and they can become exposed to factors even without having to name them as such. Prime numbers are important to understand in mathematics, and this activity allows learners to informally play with the concept of prime without having to memorize a definition. Like many of our activities I hope that the work I do now will provide solid groundwork for their future math work.

Briefly tell the story of the ancient Greek mathematician:

- – Was a Greek scholar of the third century BC.
- – Credited with many discoveries, including the measurement of the circumference of Earth.

–   Created a simple algorithm for determining prime numbers.
–   Was director of the famous Library of Alexandria
–   Was highly regarded and respected, but not much of his writing survived.
–   Starved himself to death because he was so upset he had gone blind.

Explain that numbers have interesting characteristics about them, as the children know, but that they do share similarities and some are more alike than others. Some are unique. Talk about what a sieve actually is, acting like a strainer when draining pasta or cooked vegetables. We are going to use the Hundred Chart to sift through the numbers to find those of particular importance that have a similar characteristic.

Talk about one type of number that is used to identify or name a number. Prime is the name given to numbers that can be broken only into even groups of the number and one.

Can mention that we call these factors but not really necessary. For example, 7 can only be made with 1 group of 7 or 7 groups of 1, while 4 can be made by 2 groups of 2 as well as 1 group of 4 or 4 groups of 1. Therefore 7 is considered a prime number. Explain that mathematicians find prime numbers particularly interesting and that they have been studied for a very long time and continue to fascinate mathematicians. To find all the prime numbers up to 100 we can use Eratosthenes's idea.

Ask the children to pay attention to things they notice. Ask them to look for patterns, relationships as we create the sieve. I could explain that 1 is not a prime because a prime has 2 factors exactly (1 and the number). We begin by crossing out 1 with an X. Ask the children to circle the next number, 2. Then they need to count by 2's and cross out each number they land on all the way up to 100. Then they circle the next number not crossed out (3) and, counting by 3's, cross out all the numbers as they land on them. Then they go to the next number not crossed out, and circle it (5) and count by 5's, crossing out all the numbers they land on. Ask the children to continue to do this and when there are no numbers left to cross out, then all that remains are prime numbers.

Allow anyone who so wishes to work with a partner. Have a brief class conversation after they finish about what they noticed. Note their comments on chart paper.

Next, hand out fresh Hundred Charts. This time, ask students to get out their crayons and use a different color for each number that they are counting by. Circle 2 and then ask them to use a blue crayon as they count by 2's and color each square blue that they land on as they count. Then circle 3 and using a yellow crayon, color each square yellow as they count by 3's and so on. This will bring out some interesting relationships that the first round of the sieve might not evoke. Discuss how certain numbers have more than one color. Look at those numbers. Discuss in whole conversation, noting on chart paper their ideas. Ask them to explain their thinking and work in their math journals. They can staple their Hundred Chart in their journal too.

Post-Lesson Reflection

Funny how things never go as you think they might. The children surprise me, no matter what we are doing. One of the things Jonathon noticed was that some of the primes were close together but most were far apart. I ended up talking about 'twin primes' and how that is an area of number theory that is thought about by mathematicians. When I told them that the next twin prime after 41 and 43 was 821 and 823, they thought that was crazy. Sophie said she was going to sift out all the primes between 100 and 200 tonight. I told her she could report back to us tomorrow what she found out. Karen was right. This was a great way to expose the children to another area of number theory. I cannot believe how excited they were to do this. Having a mentor teacher like Karen has been so helpful. I am so grateful. I need to send her an email and let her know how today's lesson went and I have to update the class web site and put this activity on our math page. There is much to be done!

(Back to A Foreign Land)

A small dark-haired girl came through the doorway. "Excuuuuuse meeeee!" she said to the guests. The visitors turned around as she tried to make her way through the group. She was moving with purpose. One of the visitors jumped back immediately when he saw what was in the girl's hand.

"I have to feed our snake," the girl said matter-of-factly. In the palm of her hand thrust out before her, secure in a sandwich baggie, rested a tiny mouse, freshly thawed and just recently housed in the freezer section of the refrigerator of the staff room. With apparent interest, the visitors followed the girl as she made her way to the terrarium in the farthest corner of the room. "We have a corn snake," she explained, "He was born in captivity and we adopted him. His name is Elaphe, which is Latin for 'corn snake.' Do you want to watch him have his lunch?" Within seconds the visitors witnessed the orange albino snake's jaws dislocate to accommodate the offering from the hand of the child. Another girl wiggled between the visitors to see the snake eat its lunch. Without turning to face the adults, she chimed in, "We buy all the mice with the money we made from our craft sale!"

Craft Sale: Empowering the Children

The snake did not just appear in the classroom but became our class responsibility after a great deal of work by the children. What might appear as a typical classroom pet was actually the result of an empowering experience in which the children were completely responsible for all aspects of the snake's existence. It was also a deeply integrated and important curricular event. Let me explain.

During the grade-two teaching year, I had the great fortune to share some of my classroom responsibilities with another teacher, Cheryl. Very early in the school year, a parent had donated a terrarium to us, so together we took the question to the children. "What might we do with the terrarium?" This question turned into a powerful project during which the children began to research all aspects of housing an animal in the classroom. They phoned veterinarians and asked questions about longevity, dander, hardiness, food, and breeding in captivity. They phoned pet stores, animal breeders, and reptilians. They searched web sites, read books, and talked to current pet owners. They divided themselves into groups according to the type of animal they felt was most appropriate. Their task was to research as thoroughly as possible all aspects that must be considered when acquiring an animal. Then, they had to make the best argument possible to the rest of the class through a prepared presentation in whatever form they wished. Some created scripts and performed their presentation. Others staged puppet shows. Still others did straightforward PowerPoint presentations.

It is important to note that while I was intentional in using technology in meaningful ways in all areas of the curriculum, with this particular project I had not asked the children to consider the use of technology. They could make their claim any way that worked for them. In every single case they used technology to help them make decisions, to communicate with other people through emails, and to do research. At one point during this project I had a researcher visiting the classroom who was interested in children's engagement with technology. I suggested he hang out with William, since I thought that William was working on a PowerPoint presentation. Within minutes of sending the university scholar off with William, the researcher returned, confused and frustrated. I asked him what the issue was. He said that William was not creating a PowerPoint presentation at all; rather, he was making a puppet stage. Sure enough, I found William around the corner in the art room with a parent assistant and some other children. He was definitely creating a puppet theater out of a cardboard box. This is how my conversation with him went:

MARY.   William, what are you up to?
WILLIAM.   I am making a puppet theater for Hammy.
MARY.   Hammy?
WILLIAM.   Yeah! Hammy! My hamster puppet!
MARY.   So you are doing a puppet show for your presentation?
WILLIAM.   Yep.
MARY.   Okay. But I am a little confused. I thought you were doing a PowerPoint presentation.
WILLIAM.   *I'm* not doing a PowerPoint presentation, Ms. S. *Hammy* is!

It had never occurred to me before that to William, the PowerPoint was only a part of a puppet show and the technology was going to be used by the puppet, who was pushing the idea of getting a hamster for the classroom. The researcher was a little speechless as he observed what to him would be considered foreign territory. He was fascinated to witness the way the "digital natives" (Prensky, 2001) of the classroom were thinking as they lived with technology in our place of grade two.

In the end, the group who argued for a snake made the best case, and they won over the rest of the class. Then the search began for a snake that was in need of adoption and who was born in captivity. Part of the research into a snake was the cost of the adoption as well as the ongoing cost of housing and feeding the snake for its lifetime. Together we added up the costs. My teaching partner, Cheryl, and I asked the children how we might raise money. I had been thinking we could sell bags of popcorn. However, the children felt strongly about putting on a craft sale for the school. They seemed to understand the amount of effort that would be needed. Cheryl and I had to stop ourselves from shutting down this possibility when it was first proposed because of the scope of what the children were proposing. Why not have them create a craft sale for the school and community? My teaching partner had experience with craft sales, and soon the children were working hard at creating patterned necklaces and bracelets. Some children took their work home and made birdhouses. Others wrote short stories. Some created art and framed it. Many of the children baked cookies and gumdrop cakes at home, carefully measuring ingredients and following directions. Before I knew it, the children had created all that they felt was needed.

The next task was to decide how much to charge for each item. It was such a wonderful problem. We had conversations around money, around value, and around worth. We talked about what was too much to charge, and examples of charging too little. Eventually the children negotiated prices for everything and with Cheryl's guidance affixed all the tags. Every child had a job at the fair. Some were greeters who explained to shoppers where to go and how to proceed. They passed out baskets for the shoppers who wished to collect items. Other children were charged with making change for the purchases, while being supervised by grade eight students and a few parent helpers. Still other children wrapped up the goods after they were purchased. It was a huge success and after the dust settled, the children had raised more than $400 to go toward the care of the classroom snake. With the help of a provincial reptilian society an appropriate animal was found and, as the child in the main narrative explained, Elaphe became a part of our class. He had provided the children with the opportunity to see that they could very well accomplish something challenging. He provided for us, the two teachers, an excellent opportunity to meet curriculum expectations while working in a meaningful context. We could have set up opportunities to play with fake money, to create a make-believe store, for example, but I am grateful for the real learning experience the craft sale provided for all of us. It was all thanks to the children's ability to imagine what was possible.

(Back to A Foreign Land)

The grade-two student turned and looked at the visitors. "Snakes poop too! Did you know?" she said with a toothy grin. The feeder of the snake continued, "I have to clean that up next." She paused and then added, quietly giggling, "You can help if you want!"

"Isabel, that's YOUR job!" For the first time, the visitors heard a distinct adult voice. Across the room, a woman stood up from a small group of children who had been huddling around one of the Lego robots.

"Hi. I'm Mary. I see you've met Isabel. Welcome to our grade-two class!" Making her way around groups of children, the classroom teacher greeted the visitors by shaking their hands. Out of thin air, a brown-haired boy appeared at the teacher's elbow.

"Excuse me, Ms. S, but we need clarification. Does it have to come back to the *same* spot?" Smiling, the teacher gestured to the visitors to accompany her to the electronic whiteboard where three children were working with the brown-haired boy named Jordan.

"Well, Jordan, what does the question say?"

Jordan looked up and read a problem off one of the sheets of paper on the retractable wall:

Grade Twos, Your task for today is to create a robot that goes forward for 2 meters, turns around, and comes back to the place where it started. You can do it! Ms. S

"So, what do you think it could mean?" the teacher asked.

"I would say that means the same spot. Right guys?" Jordan turned to the rest of his group and continued, "Let's just do it based on that thinking."

"That's what I told you we should do, Jordan!" the other boy in the group said. "Come on! We're wasting time! Sophie already has the program begun on the SMARTBoard and she's waiting for us! I wanna figure this out." The red-haired girl, Sophie, had a program open on the computer that was attached to the SMARTBoard. Immediately the guests noticed that the electronic whiteboard was now transformed into a very large computer screen and the children were tapping it with their hands. Items appeared, disappeared, and were moved around as their fingers dragged icons all over the whiteboard. There was great negotiation between them concerning which motors to turn on, in what direction, and for how long, and in what sequence the events should occur. To the visitors it might have seemed as if the children were simply linking together colorful puzzle pieces.

"We're programming our robot," Jordan said matter-of-factly to the guests when he noticed them watching. "It's not that easy a problem, but then again, Ms. S doesn't give us easy problems *ever*, right, Ian?"

"Yeah!" added Ian, who had been so impatient to get started moments earlier. "Ms. S says the best problems are the ones that are hard because then we have to think! Remember when Dr. Karen and Ms. S had us counting grains of sand? That was crazy hard but we figured it out!"

Walking Beside My Mentor: Counting Grains of Sand

Karen, a mentor teacher who came to my class once every few weeks, was experienced in the Gadamerian sense. She could see possibilities in the actions and words of the children and in mathematics. We had worked together previously as teachers on the same staff. I trusted her, and she trusted me. During my grade-two year she came many times to work with me and the children as we made our way through mathematics. The children began to look forward to Karen's visits and I embraced the opportunity to learn right along with her in our team-teaching opportunities. During her visits I encouraged parents to come and learn with us. It is worth noting that the mathematics we took up was not always directly connected to our bigger inquiry work. Indeed, there were multiple relationships between what we did, but in the following example, counting grains of sand was not part of our larger Martian exploration. Mathematics itself is worthy of standing on its own. About this, Stewart (2006) notes:

> Math done "for its own sake" can be exquisitely beautiful and elegant. Not the "sums" we all do in school; as individuals those are mostly ugly and formless, although the general principles that govern them have their own kind of beauty. It is the ideas, the generalities, the sudden flashes of insight. (p. 9)

I recall how vibrant and alive the class was on the morning we counted grains of sand. Indeed, there were sudden flashes of insight, and perhaps the greatest came from a parent. The following narrative is but one example of work we undertook together.

ACT 1.	Scene 1.
	[Grade-two classroom. Twenty-eight children sit around a homemade round table from grade one and painted in grade two. The blue legless table is divided into twelve equal sections with each section containing one of the symbolic representations of the twelve signs of the zodiac. A bright yellow sun is painted in the center of the table. A guest who is a mathematician and a mentoring teacher sits at the table with the children. A parent, Marlie, joins the children at the invitation from the classroom teacher, Mary, who also sits at the table with a marker and chart paper to scribe the conversation and to learn from the mentor teacher, Karen. The parent has not been given a task to do other than to engage in the same mathematics as the children. Karen takes a coffee filter filled with beach sand and sets it in front of her on the table so the children can all see it. Then she speaks.]
KAREN	[Mentor Teacher]. So how do you think we can count sand?
STUDENT.	We can make a machine, dump the sand on it, and estimate how much sand is on it from the weight!
MARY	[Classroom Teacher]. If I know how much this weighs, would I know how much the rest weighs? You're on to a good idea.
STUDENT.	You can ask another person, maybe.
KAREN:	Hmmm. You must be able to rely on them. How do you know whether people you ask are right?
STUDENT.	Well, if you asked Einstein he might know!
STUDENT.	You can get a counting machine of some kind.
STUDENT.	You can pile the sand on each constellation picture on our round table and count the sand on one of the places.

KAREN. Uh huh! Let's say there were one billion grains of sand. So then, on each constellation there were one billion, twelve times. So there would be...

STUDENTS. Twelve billion!

KAREN. So by knowing one you would know all.

STUDENT. But what if they are different sizes?

MARY. Uh huh! You have to figure out how to make them all the same size.

STUDENT. Does it have something to do with time?

STUDENT. A sand timer any help?

KAREN. So now you are thinking about using sand to count time.

STUDENT. We could use a scanner or something. Put the sand on it and the machine will scan it to see how many grains of sand there are.

KAREN. Scientists have made a sensitive type of board that shoots an atom with a light that smashes into the atom and counts all the little pieces.

STUDENT. That would be a great science project.

MARY. That would be a really big science project!

STUDENT. It will be different if the sand was wet.

STUDENT. Yeah, I know. Wet sand sticks to my fingers. When I make castles at my grandma's cabin in the summers I need to use the wet sand.

STUDENT. Well, if I can take a pinch of sand and I could count one pinch, could I count the rest in pinches?

KAREN. Does that sound good?

STUDENT. Not if it was an entire beach!

STUDENT. You could count the pinch in buckets. And then buckets on the beach!

KAREN. Let's think about this idea. Let's say we put twelve equal piles of sand on each of these signs of the zodiac on the table. You said before that if you knew there were a billion grains of sand sitting here on, say, the Scorpion, [points to the zodiac symbol to the right of her on the table] and you know for sure that the amounts are pretty equal, you could tell how much was on the whole table?

STUDENTS. Yup! Twelve billion!

KAREN. So by knowing *one*, you know all? When I said I would take one grain, take my idea of one, and then take that to a pinch, and that...

STUDENTS. to a handful!

KAREN. to a...

STUDENTS. bucket!

KAREN. to a...

STUDENTS. Truckload!

KAREN. What do we mean when we say *one*? Work as partners or alone. Figure out a way to count the grains of sand. You are actually going to do the counting. Ms. S is going to bring you a coffee filter of sand once you have found a suitable spot to think and to work.

[Fade to black.]

ACT 1.   Scene 2.

            [Children are spread out in the classroom. Some have chosen to work on
            their own, but most work with a partner. Karen and the teacher, Ms. S,
            are moving between groups, asking questions, paying attention to decisions
            students are making. The parent is squatted down and working on her own
            at a desk, counting and concentrating. She has one hand running through
            her hair as she is thinking. Classroom teacher approaches.]

MARY.   How's it going, Marlie?

MARLIE   [Parent]. *[smiling]* Math wasn't like this for me twenty-five years ago!
            This isn't exactly easy, but I want to figure it out! I can't imagine being
            bored to death if I had gotten to think about counting and numbers
            stuff like this!

MARY.   Do you think you have a good sense of numbers? Can you tell by look-
            ing at your pile about how much might be in the coffee filter? Do you
            remember ever talking together as a class about how you figured out
            problems?

MARLIE.   Are you kidding? We corrected our problems together and that was it.
            If I had a different answer, I had to put in the right one. I never ever
            talked about why I had a different answer or was asked to figure out
            what I might have done that led me to a wrong answer. It was either
            right or wrong. I knew very well how to carry the one, and add up
            rows, but I don't think I ever had the chance to make much sense of
            the size of numbers. I followed the rules. That was it. And I was a very
            good rule follower. *[Marlie rolls eyes.]* I just was never asked to actually
            think much when it came to mathematics. I really don't want Morgan
            to experience math the way I did. I mean, right now, you are asking me
            to think and figure out how to do this. I am actually having to think.
            [Student with brown ponytail bounces up to parent.]

STUDENT.   Hey Mom! Do you have it all counted yet? Are you having fun in
            grade two? Me and Jordan got 124 grains in one pinch, but your fingers
            are fatter than mine. So what did you get?

MARLIE.   Morgan, I think I just might need your help.
[Fade to black.]

ACT 1.   Scene 3.

            [Back at the round table. Chart paper sits to the right of Karen, behind some
            children who are all around the table. On the chart are written some results:
            Jordan: 8492 grains. Sophie: 11,500 grains. Hannah: 10,692 grains. Ian:
            129 grains in one pinch.]

KAREN.   What did you learn?

STUDENT.   We got about 100 grains in one pinch. Another pinch the same size
            we automatically knew it was about 100.

KAREN.   Okay. So what did you do? You broke the whole idea apart by figuring
            out a part of it. That's what mathematicians do all the time with big
            problems. They never figure out the whole thing. They break it apart!

STUDENT.   It was way too hard before that!

KAREN.   Like Alice in Wonderland and the Queen of Hearts. The Queen of Hearts said something like: "Do you know math? What's 1 + 1 + 1 + 1 + 1…" and Alice said, "Stop. You are making my head hurt!" I think the Red Heart said something like, "See, she can't do math!" Poor Alice! Did you know that 1 + 1 + 1 + 1 + 1 + 1… is what we used to have for counting? Early people figured out that was way too hard, so they made another way.

STUDENT.   So they had, like, one and another one and another one and another one instead of 1, 2, 3, 4, 5, 6, 7…

KAREN.   Yup!

STUDENT.   I am so glad we got numbers now!

KAREN.   Me too. Your group, Hannah, used a bottle cap. Can you explain why?

STUDENT.   Well, we knew that a bottle cap always stays the same size so it should be the same amount.

STUDENT.   Yeah. We had to make sure our pinches were the same size too.

KAREN.   You used the bottle cap as a standard. That's interesting.

STUDENT.   We had 81 bottle caps in one pile and we had twelve piles in one group.

KAREN.   How much in one of your groups?

STUDENT.   Twelve. So I added 81 twelve times. I got 972 grains.
            [*Student shows on chart paper how she added up the 81 twelve times.*]

STUDENT.   That makes sense.

STUDENT.   And then we had 11 groups. And since 972 is really close to 1,000, we figured our answer would be close to 11,000. We used a calculator and punched 972 then x and equals. The answer was 10,692.

STUDENT.   We did it differently than that.

MARY.   What did you do, Isabel?

STUDENT.   Well Brent and I had 80 grains in one pinch.

KAREN.   I remember seeing how you organized your work. Can you explain?

STUDENT.   For every pinch in our total amount we put a mark, like a dash, like this [*goes to chart paper and makes a downward dash, /*] and we grouped them in fives. [*Turns back to chart paper and makes tally marks in groups of five.*] So then when we were out of pinches, we counted up the marks by counting by fives and we got 165 marks.

KAREN.   And then you had to add 80, 165 times and people call that multiplying. So we knew 80 + 80, that would be 2 groups of 80. About how many are 165 groups of 80?

STUDENT.   80 is close to 100, so it might be like 165 hundred!

KAREN.   We write it like this: 16,500, and we read it like this: sixteen thousand five hundred. [*Karen motions her hand across the number as she reads it.*]

STUDENT.   But it's gotta be less than that, because we added an extra 20 on to the answer 165 times.

MARY.   Maybe you can explain what you mean in your math journal, Ryan. Perhaps now would be a good time for everyone to find your

math journals and explain what you did this morning. Include any
frustrations you had but please try your best to include the strategies
you used to overcome those frustrations!

[Fade to black.]

ACT 1.   Scene 4.
                *[Parking lot in front of the school. Classroom teacher is walking parent to
                her car.]*

MARLIE.   You know, my head is spinning a little. I guess I didn't realize that
                when you said I could come in and do math with the children that
                I'd actually be doing math. I assumed I would be pretty bored to death
                with the ease of the work. But I had to think!

MARY.   I am really glad you took up our invitation to come today, Marlie. And
                I am glad your head hurts a little! I want the children to do things that
                make them think, that help them understand that math is something
                that people shape and continue to shape.

MARLIE.   I guess I never thought about counting before. I never thought about
                where numbers came from because I just thought they always were.

MARY.   The problem we worked on today is an example of a problem that
                helps the children have a better understanding of numbers, counting,
                estimation, and having our mathematical world make sense. One of
                the things that Karen and I try to bring to the children is the no-
                tion of mathematics making sense and that it comes from somewhere.
                I have learned so much from working with Karen. The children love
                when she comes to our class because they know she will challenge
                them with something interesting. It gives me a chance to listen to
                how she is with the children, to become exposed to new mathematical
                understandings, and to pay attention to the children in a different way.
                I'll put on our class web site when she will be returning. I hope you
                can join us again!

MARLIE.   I will try, for sure. I do have a question. In your curriculum night
                for parents you spoke about the importance of integrating different
                subject areas of the curriculum into topics of study and that your Mars
                exploration is an example of that, but I was wondering, How is count-
                ing sand connected to that project?

MARY.   That is a terrific question. You see, while the idea of patterns and
                relationships and number sense is part of much of what we do, this
                particular task was not directly connected. It is not necessary to make
                everything be connected. Sometimes the mathematics just stands on
                its own as an interesting problem. It doesn't make sense to force con-
                nections that are not authentic. For example the work we are doing
                building the Martian terrain has a lot to do with ratio and scale so
                we will work on those mathematical ideas while we build the terrain.
                But thinking about counting, and the many relationships that were

involved in our work today, is not directly connected to our inquiry about the colonization of Mars.

[*Bell rings to indicate time for children to come inside from lunch.*]

MARY.   That is my cue! I really appreciate your interest in our work, Marlie. Morgan will be taking home her math journal tomorrow. Let me know whether you have any questions about her work or problems that we have been doing. Thanks for coming. Let's keep talking about this. You are always welcome to join us!

[*Fade to black.*]

(Back to A Foreign Land)

Jordan turned to the guests, eyeing them up and down. Then he spoke: "We get lots of visitors you know. It's kinda neat, because we get to meet people from different countries and they sign our online guestbook on our class web site and email us when they go back home and we write back to them. You should sign it!"

Ian chimed in, excitedly, "Sometimes people don't speak English and they bring transporters with them!" "Translators, Ian," corrected Sophie, the red-headed girl who had remained quiet until now. She turned to the visitors: "Men in suits came from Sony in Japan to learn from us last week! What languages do you speak? Do you need a translator? Where did you come from? Why are you here?"

"Sophie, you will have time to talk to our guests later on today after lunch," said the teacher, "I am sure they would love to answer your questions then, okay?"

"Sure, Ms. S!" Facing the visitors she added, "Talk to you later!" and turned on her heel. Looking up at the SMARTBoard she seemed to be reading the program's sequence to herself and then shouted, "Don't forget! We have to turn *that* motor [*she points at one of the colorful bricks on the SMARTBoard*] off at the end!"

"Oh! You're right!" said Jordan. "Do it, please, Ian, and let's transfer the program to the RCX box now and see what happens!"

"Grade 2S, can I have you stop and listen just for a moment?" the teacher said in a voice just loud enough for the farthest group in the room to hear. The buzz in the classroom simmered and then quieted as the students learned that the teacher wanted to tell them something.

"At 11:30 we are going to come back together to discuss what we've discovered in our work this morning. We will be talking about our *learnings*, *frustrations*, and *strategies* as always, so please make your way to the Martian terrain at 11:30 for our class meeting and take any notes you might have with you. Please pass this on to the group working in the conference room, Brent. Sorry for the interruption everyone!" Then the teacher vanished into a group of children once more. The visitors noticed Ms. Rose beckoning them at the doorway. They made their way to the hall.

"I know it is difficult to talk much while class is going on," she said to the group, "but I have a suggestion for you. I am going to fill in for Ms. Stordy while the children eat their lunch, so you can talk with her and ask any questions you might have then." The group members smiled and nodded at this suggestion. From this response it seemed as if they had many questions. Eager to return to the room, the group headed back inside, except for one guest who had a furrowed brow. Noticing this, Ms. Rose approached him.

"Are you okay?"

Hesitating, he seemed to choose his words carefully. Then he spoke.

"Look. I've read about inquiry-based learning and solving ill-defined problems, but still you should know that I am concerned with how you know what kids know when you work like this. I mean, where are the math textbooks? Do they do Mad Minutes? There's got to be a balance. These kids can program a robot, but can they add two simple numbers together? Working like this is fine when you don't have any children with learning challenges. How can children focus in a setting like this? Everyone knows that children can only concentrate for a short amount of time. They simply don't have the attention span. There can't be any children with attention difficulties in that room, right? They would not be able to handle it."

Challenging the Problem with Attentiveness

[Fade in.

Int—grade two classroom—day]

[Afternoon sunlight shines in through the window with tall trees swaying wildly in the background from the warm Chinook winds. The classroom is buzzing with activity. The atmosphere is relaxed. Upbeat dance music is playing on the CD player. Twenty-three children and a teacher are crowded around the Martian terrain, which is positioned in the center of the room. They are painting the scaled-down version of a section of the planet Mars with reddish-orange paint. They are wearing oversized long-sleeved, light-blue men's dress shirts that are buttoned up the back to protect their clothes. The painters are on their feet, reaching for fresh paint in old ice cream containers and moving to the beat of the dance music. Some children are singing along with the music. Four more children are huddled together on the floor to the left of the terrain, making adjustments to a robot's design. One boy works alone at one of the classroom computers.

Two researchers, a man and a woman from two Canadian institutions, are sitting on chairs on the perimeter of the classroom, watching the class activities. They are watching the children and teacher and every now and then write something down in a notebook. Camera slowly scans the entire space and rests on a view of the whole room.]

NARRATOR (TEACHER). [V.O.] It was such an enjoyable spring after-noon. It was time to paint the terrain, and the children were excited about doing it. Finally, our piece of Mars was beginning to shape up, and painting it red was one of the final touches before we could begin to use the terrain with the robots. This was to be an easy afternoon. The children had been working hard at their research and on designing and programming their team robots so I felt that a low-key afternoon of painting was in order. Everyone seemed to be in a good mood. As I painted with the children I glanced around the room, as teachers do, to ensure all students were safe and doing something acceptable according to our class guidelines. The room was noisy. The paint-ers were chatting and singing as they worked and the small group on the floor had their heads together, discussing the best position for their light and touch sensors on their robot. Ian was working alone. He was standing at a computer and from where I stood I could see he had the Lego Mindstorms programming software running.

[Camera moves to the hands of the clock on the wall. It is a quarter after one.]
[Fade out.]
[Fade in.]
[Camera is still on the clock; however, the time now reads three o'clock. Camera zooms out to the whole classroom once more. The terrain has been completely painted red. Teacher enters the classroom with an ice cream container of clean wet paintbrushes. Two researchers are standing and facing the computer where Ian had been working earlier. No children are visible in the classroom. One researcher turns her head when she hears the teacher enter the room. Their eyes meet. The researcher motions with her eyes downward, for the teacher's eyes to follow. There, behind the researchers stood Ian, who was still working away. Teach-er's eyes widen with surprise.]

TEACHER.	Ian! You have to get ready to go, or you will miss the bus! Everyone else is gone!
	[Little boy turns, smiling and excited.]
IAN.	Okay, Ms. S! I think I figured it out though! We have to paint a black line all around the outside of the terrain! I have a program created now that I think is going to work!
TEACHER.	Awesome! We will need to talk about it first thing in the morning at our class meeting, okay? But right now

you have got to get your Go-Go-Gadget legs moving and make that bus! I'll log you out and save your latest programming to your F drive. You just get moving, mister!

[Boy turns, and almost runs into male researcher as he makes his way to his backpack and jacket. He whips off his indoor shoes and chucks them on top of his desk. One falls to the floor. He bends to get it.]

TEACHER. Never mind, Ian. I'll take care of your shoes. You go! And if you do miss your bus, it's not the end of the world. You'll get home. You know that. But I need you to try to make the bus first!

IAN. K! Thanks Ms. S! See you in the morning!

[Boy races out of the classroom. Teacher collects his shoes and straightened up papers on his desk and walks to the computer to save Ian's work. Next to the computer sits a robot five inches from an infrared light transmitter that is attached to the computer. The robot sits on top of an old, hardback textbook. Teacher points at the book and laughs.]

TEACHER. See? I guess we DO use textbooks in this class!

[Researchers chuckle politely.]

MALE RESEARCHER. What's the story with the textbook? Why is it there?

TEACHER. I grabbed that book out of a box of old books that another colleague donated to me. Ian must have needed something thick enough to position his robot to the right height for the transmitter's infrared light to reach the appropriate spot on the RCX box on the robot. That is how the children get the program from the computer to the robot. They transfer it through infrared light, sort of like how many remote controls work for televisions.

FEMALE RESEARCHER. That's pretty decent problem solving if you ask me.

TEACHER. Well, Ian is a great problem solver if he is doing something he deems worthwhile. One of our earliest learnings with regard to our work with the robotics was that light travels in a straight line, unless it gets refracted. We had such a great chat about the kinds of light that actually exist even if we can't see it. This book, positioned like it is, is very telling that Ian remembered that idea. I try to write down and pin up the things we figure out in our work. That way the children can build on the knowledge of each other.

MALE RESEARCHER. I've been hanging out with these children for a while. It seems to me they work in a way that is truly collaborative, rather than competing with one another.

TEACHER. I love that you noticed that. I couldn't ask for anything more. The buzzwords might be that we are a 'knowledge-building community' in this class, but actually it is true! I have seen many instances of how the children help each other and give strategies to one another. They want one another to succeed.

FEMALE RESEARCHER. What's the story with Ian?

TEACHER. What do you mean?

FEMALE RESEARCHER. Well, I've been spending a great deal of time thinking about engagement, and what it means to be engaged. We want children to be actively engaged but we don't do a good job of explaining what that means. Today, I sat and I observed Ian for two hours. It was fascinating. His actions today are the epitome for me of what it means to be truly focused, to be completely engaged.

TEACHER. Ian was in grade one with me last year. I love him to bits. He has a lot of energy, and he landed in my classroom last year full of exuberance for everything. When I met with his parents very early in the grade one year, they asked me to do one thing for their son. They asked that I do whatever I could to ensure he never loses his enthusiasm for learning.

MALE RESEARCHER. So how do you do that, as a teacher?

TEACHER. I believe my job is to do that for all my students, but for Ian, I could tell it was of utmost importance. As a student he is a bundle of energy, as are many of my students. However, Ian's energy level is very high. He cannot really sit still, so I don't ask him to do that very much. I know children diagnosed with attention deficit disorder who have similar attributes as Ian. I don't believe it makes sense to expect young children to remain still for long periods of time. I find it hard to do that myself! Sometimes I need to move in order to think. Why would children be any different? Last year a visitor from Texas was hanging out with my class and I noticed she was talking to Ian and had tears in her eyes.

When I saw this I went over immediately and wondered what Ian could possibly have done. He can be a bit of a monkey, and I mean this in the best possible way, so it would not have been a surprise to hear he might have done something slightly silly. The lady turned to me and said, "I have a young boy at home who is exactly like Ian and it just makes me so happy to see that there are schools where the Ians of the world are allowed to just be themselves. You have made a space for them. Thank you." I got a little choked up when she said that to me. I still get choked up, actually.

[Teacher puts left hand to her left eye and wipes the corner of it with her finger.]

FEMALE RESEARCHER. That is so interesting! I was fascinated watching him. He worked on the computer making adjustments to his program. Then he would download the program to his robot and march out to the hall and try it. After he observed the robot's actions he would pick it up, squeeze past all of you painting and continue to make changes to the program at the computer. He must have repeated the process of adjusting and testing his work half a dozen times at least.

MALE RESEARCHER. I've been intrigued by Ian for months now, ever since I started research in the school. When he is interested in the work, when he has a say in what he does with his time, he is completely focused, like you witnessed today. I get so frustrated when I overhear educators say children can only focus for short periods of time so they chop everything up in the curriculum into tiny bits. Ian disproved that theory in his actions today. While he was solving his problem, of figuring out how to keep the robot on the terrain was it?

TEACHER. Yes.

MALE RESEARCHER. Right. Well, when he was doing that, he was working like a mathematician would. I don't mean so much trial and error, which is very much a mathematical way of working, but perseverance. He was narrowing down the possibilities in the logic of his program as he tried to find a solution. From what I know of the Mindstorm Lego programming, it is all about problem solving and logic. You can't have a wheel turn if a motor is not turned on. You can't have a motor work if the power has not been told to turn on. You cannot have a motor stop in time if you haven't told the light or touch sensors what do when encountering certain elements.

TEACHER. I know that he was interested in solving our issue of how we are going to keep our robots on the terrain in the first place.

FEMALE RESEARCHER. Sounds like he figured it out!

TEACHER. We were sitting around the terrain the other day, no, it was yesterday, and one of the children asked how we would keep all the robots from falling off, since, as you can see, the terrain is about six inches high all around the edges. Honestly, I told them I really didn't know but that perhaps we could figure it out together. So it seems this is the problem that Ian was working on all afternoon. I hadn't even gotten to think about the possibilities myself!

MALE RESEARCHER. Well, if you don't mind, I'd like to sit in on tomorrow morning's class meeting when Ian shares what he figured out.

FEMALE RESEARCHER. I wish I could be there too, but I have to teach tomorrow at the university. I just wanted you to know how it blew me away that Ian could be that focused considering all the 'distractions' around him. I mean, you guys were singing with the music and dancing around as you painted. This room was not quiet. Anyone who would suggest Ian has an attention disorder needs to hear what he was able to accomplish today.

TEACHER. If the situation were different, if I insisted that Ian sit still at a desk all day long and do work out of a math textbook, and if I gave him pages and pages of computations and other work over which he didn't have any control, I would say he would be a very different student. Slowly I would see his spirit lose its glow. He would become someone else. The thing is he loves mathematics so I want to make sure he gets the chance to actually think mathematically and do interesting mathematical activities. For two weeks after we did Martian Math, Ian insisted that Earth math (base ten) was much too easy and wanted to solve all problems using Martian math (base five). I had told the class a story about Martians needing to count a big pile of rocks up on Mars but that they only had the symbols 0, 1, 2, 3, and 4 to work with. I asked them to figure out how to count a pile of rocks. Place value is so critical in grade two. Working in other bases allows children to see that there is a pattern to our numbers. It wasn't until we worked with base five that some of the children really got a deeper understanding of the importance of the positions of the digits. It's such a great problem. I learned about it from Karen. Ian loved it.

FEMALE RESEARCHER. There is no question in my mind that many of the disenfranchised and bored children who end up getting in trouble and being sent out of the classroom by their teachers are most likely learners like Ian. Isn't it Maxine Greene who says we need to create spaces for the bored, the marginalized, the disenfranchised? Classrooms need to be places where children are deeply engaged in meaningful worthwhile work? It seems to me Greene would say you have created a space for Ian.

TEACHER.	Let's just hope there was space for him on that bus! I better go check the office just to be sure he's gone. [*Teacher gets up and leaves the room. You can hear her footsteps in the hallway growing dimmer and dimmer. Male researcher picks up heavy textbook and smiles. He turns to Female Researcher.*]
MALE RESEARCHER.	Finally, a really good reason for a textbook in a grade-two classroom. Great line. I need to use that one myself. [*Female Researcher nods her head and gives a little giggle.*]
FEMALE RESEARCHER.	Me too! Tomorrow morning with my BEd students! I'll beat you to it.

[*Fade to black.*]

(Back to A Foreign Land)

The man in the hallway with Ms. Rose continued with his questions. "As the principal, aren't you worried? How do you know your teachers in classrooms like this one are meeting the curriculum?"

"Well, those are interesting questions. I encourage you to ask Ms. Stordy exactly what you've asked me when you get together at lunch, but in the meantime, I suggest you go back to the classroom and start to listen to the children. Ask them about their work. Listen to their class conversation about their learnings, frustrations, and strategies from this morning's tasks. Since mathematics is important to you, I also suggest you ask a few of the children to show you their math journals. I'd go so far as to argue that a lot of the students have a better sense of numbers than many adults. If you still have questions, please come see me before your group leaves and we can continue the conversation. But right now, I have another classroom I have to get to before lunchtime." With a parting smile, the principal hurried to the other end of the school to join a grade-four class and the lone visitor returned to the classroom to look for evidence of what the children were really learning.

Alien Territory: Treading Rough Ground

In the main narrative above, the student Sophie asked good questions of the visitors. She wondered who they were and why they were there. Her questions are hermeneutic; that is to say, they echo some of the questions of the work of hermeneutics. She was trying to place these visitors in the world. She was trying to understand how the people came to be there in her classroom, in her life-world. As the teacher in the above narrative, I also wondered these same questions of people who would come to see our classroom. When visitors would come I would want to know as much about them as possible so that I could begin to understand where they were coming from, not in the same sense that Sophie most likely wanted to know, but I wondered how familiar visitors were with making sense of what they were seeing. I wanted to know whether they had walked the terrain of empowering children to be problem solvers, to be decision makers, to reason, to think—like an engineer, like a mathematician, like an artist. Similar to the visitors from Sony in Japan who needed English translators for communication, I worried about how best to explain what I was attempting to do with the children without using language that would reduce our work to the lowest terms.

For many visitors our classroom was alien territory. Entering a foreign land such as our learning space could easily disorient visitors not yet familiar with imagining school the way they were seeing it. How do you help someone 'see' a place? How do you help unpack the messy, interwoven threads of the living space in which the children and I were engaged in creating and exploring? We were able to navigate our way around our earthly terrain of grade two because we were living it and it was the space we created together. We dwelled in that place, in the space that Isabel referred to as 'home' by the end of our second year together. As mentioned in chapter 1, to dwell according to Heidegger (1997) means to be "the way in which you are and I am, the manner in which we humans are on earth.... Dwelling itself is always a staying with things" (p. 329). In our class we attempted to stay with topics and ideas that seemed important to us.

Thinking back to those moments I must have been confusing when I attempted to explain to visitors what was happening in our classroom. In retrospect I can see that many times I was unable to engage in a true conversation, an element described earlier in chapter 3 as central to Gadamer's (1989) hermeneutics. This could be because a conversation is always about something. In a conversation both parties are talking about the same topic; there is a common language. I am reminded of an exchange I had with a school visitor who reduced our grade-one intellectual engagement surrounding life during the medieval time period to a project about fairy tales and castles. The visitor had not spent any time dwelling with us as a class at all, but saw the giant castle and spoke, based on what she might have believed to be true of all grade-one landscapes. When I spoke with her, I felt as if we were not speaking a common language and I recognize, now, that we were not talking about the same topic. We were seeing different landscapes and at the time, before my exposure to Gadamer's hermeneutics, I was unable to consider that what she was saying might be true of something; I held fast to my position rather than being open to trying to understand what she was really saying about her awareness of the landscape of grade one.

There were other moments during which I spoke with visitors who would want to know where the special needs children were, like the man talking to the principal in the narrative, because they did not believe that the children could be there in the midst of twenty-eight students. Of course they were present. The children in the classroom with special learning needs did not regularly stand out as being 'different' and struggling. The products of the work of the students were not identical, and therefore their work could not be held up to compare for first, best, worst, and last. It seemed that for some visitors this inability to do a quick superficial scan of ability grouping of children was troubling. I am reminded of visitors who wanted to know where Mars and Martian math were listed in the *Grade Two Program of Studies* specific learning outcomes. The questions posed to the principal in the narrative were questions that I had heard before. These questions were conventional markers that signified the place of grade two for many visitors treading on unfamiliar terrain.

For persons used to seeing children perform identical tasks quietly at desks, the vision of movement, of negotiating ways to see and solve a math problem, of working in a hallway to test a theory, could cause a disruption to their understanding of what constitutes structure. Indeed, the structure of our landscape was filled with messiness and difficulties. A great deal of effort went into planning the work in such a way that it was necessary to know the provincial curriculum inside and out so that connections could be made within and between disciplines. Instead of charting a path and sticking to it, we mapped landmarks into the teaching landscape and ensured that we encountered each cairn as we made our way. Long blocks of time were necessary to be able to delve deeply into topics. Every day included whole class conversations about learnings, frustrations, and strategies. These became important landmarks for the work. The children built on the knowledge of one another continuously; we were there, in the work, in that space together. We were working in the space of what Scardamalia and Bereiter (2006) call collaborative knowledge building as we built and programmed robots and researched the exploration of a nearby planet. This work required the advancement of knowledge within the community as a whole, which is true of all knowledge-building communities. The children kept individual notes about how they used their time, what they figured out, what frustrations they encountered, what strategies they used, and what they were going to do next. We did this because each person experiences his or her own learning within a collaborative knowledge-building community (Scardamalia & Bereiter, 2006). This is true of all teams that create knowledge together, play together, or build something together. Individuals cannot all play the same position at the same time; they are each a member of a very powerful team advancing knowledge and all that entails. It is no wonder then that a classroom that strives to be a collaborative knowledge-building community is going to function and look differently from a classroom in which everyone does the same thing at the same time to produce identical end products.

Counter to the notion of curriculum as landscape, Bransford et al. (2000) use the metaphor of the rutted path to describe how mathematics curriculum has been taken up traditionally:

> Vast numbers of learning objectives, each associated with pedagogical strategies, serve as mile posts along the trail mapped by texts from kindergarten to twelfth grade.... Problems are solved not by observing and responding to the natural landscape through which the mathematics curriculum passes, but by mastering time tested routines, conveniently placed along the path (National Research Council, 1990:4). (p. 138)

Mathematics curriculum structured in this way does not help students "learn their way around" mathematics (Bransford et al., 2000, p. 139). Students cannot become experienced in mathematics in the ecological sense that Gadamer (1989) refers to as knowing your way around. Jardine, Friesen, and Clifford (2003) explain it further:

> For mathematics to be deeply experienced, it must be drawn back into its suffering, its undergoing, its movement of becoming what it is, its living coming-to-presence, rather than its foreclosing being present. It is its "passing on." It is a fragile and finite and deeply human enterprise. (p. 47)

Taking up mathematics as a deeply human living enterprise is not easy. To teach in this way is to tread what Dunne (1993) calls "rough ground" in all its messiness and complexity. It is worth it, however, to help students explore and come to know the landscape of mathematics, rather than experience what Boaler (2008) describes as the distorted image of school mathematics.

People who have experienced the distorted image of mathematics in schools have traveled along a fixed rutted path (Brandsford et al., 2000) and many become lost when they do encounter the mathematical landscape. Indeed, such a landscape is alien territory. The majority of teachers of primary/elementary learners, including myself, were schooled along such a path. This poses a challenge for coming to understand our way around the terrain and to being open to the wonder, open to the possibilities that being in a classroom with children and mathematics can provide. In the next chapter I will attempt to peel back some of the layers of this challenge as my work with pre-service primary/elementary teachers leads me on in my journey to come to understand the complexities of teaching the living discipline of mathematics to children.

· 8 ·

LIVING WITH THE COVER VERSION OF MATHEMATICS

You know, I don't hate mathematics. I don't love mathematics. I don't know. I guess I just don't really care about it. And I want to. I want children to care about the work they do.

—*Kim, Elementary B Ed Student*

Keeping the Difficulty Alive

It was a busy time during the 2004 school year in Ms. Kennedy's grade-one class. From the doorway I could see that her twenty-six grade-one students were already working in clusters, dispersed throughout the crowded classroom. It was still winter outside of this suburban elementary school, so heavy coats puffed out the side-wall, expanding into the classroom and robbing Ms. Kennedy of space for her students' activities. Under the jackets on separate hooks dangled matching snow pants, half of them knocked off their pegs and slumped on the floor, resting on the heater below. Abandoned mittens were scattered under the nearest group of desks. This classroom smelled like every grade-one classroom I had ever been in during the winter. Wet snow pants drying on heaters give off a distinct odor. It is not unpleasant, but anyone who has ever spent time in such a space knows what I mean. It was 9:15 a.m. A quick scan of the room did not immediately help me locate the student teacher, Ella, with whom I was spending time today as her field experience/practicum instructor representing her university teacher education program.

This morning I could see that the children were occupied, working away at a mathematics activity. Usually when I slip into a classroom to work with student teachers during their first or second year of student teaching, there are many children who notice me as an outsider in their room and usually several will immediately approach me to inquire why I am there, to show me their work, or simply to have a chat and a break from a sometimes seemingly tedious task set before them. Today, no one approached me. They were all too busy.

I walked up to one group of boys to see what had their attention. They were playing with dice, rolling two dice and then performing a mathematical operation, in this case subtraction of two numbers. They took turns rolling and while one child rolled the other copied down the numbers. They excitedly showed me their list of equations already completed.

"Wow!" I said, "You did all that already?"

"Yup!"

At this time, Ella noticed that I was in the room. She had been working with another group of students doing the same activity. She came over.

"How are you, Ella?"

"Great! We are just doing some work on subtraction this morning. They seem to like it."

"Hey! Mary!" I felt a tug on my arm. Standing there, grinning a toothy smile was one of the boys I had just spent time with. "Look at this one! Look at this problem we got!" He was practically jumping up and down, and now his partner joined him.

"What is it, boys? What is the problem you rolled?"

"It's 100 take away 101!"

"Wow! That IS a great problem! So what are you thinking?" Together they chimed, "The answer's zero! 100 take away 101 is zero!" Right at that moment, the bell rang, and the classroom teacher, Ms. Kennedy, bustled over next to me to urge the children into a straight line for physical education class. The boys were practically bursting with excitement, and they needed to tell their teacher immediately about their great problem.

"Teacher! Teacher! We got 100 take away 101! And the answer is zero!" Ms. Kennedy, dealing with the business of the morning, in this case getting everyone lined up to head to the gymnasium, simply looked at the boys and said, "Line up for gym."

I remember standing there in the classroom with the student teacher beside me. This was one of those opportunities in which a pedagogic moment presented itself, rife with possibilities, rife with openings into a very interesting mathematical idea. Coming to understand is about recognizing when such moments present themselves (Gadamer, 1989). The boys' claim—101 – 100 = 0—was an opening into something mathematically interesting. Their claim hung in the air. The invitation into the fascinating world of zero, particularly zero as a dangerous idea (Enzensberger, 1997; Seife, 2000) was unanswered. The bloodlines of zero were lost. This is but one example of the kinds of openings into interesting and engaging mathematics that present themselves in many classrooms. The reasons this teacher did not acknowledge the opening is not what I am concerned with. She may have made a decision not to discuss negative numbers. She might not have heard what the boys said or she might not have recognized the possibilities that existed in the boys' statement. For whatever the reason, my intention is not to point out a teacher's inattention or lack of action. Rather, I was struck by what the boys said. I still wonder about that particular moment, but I share the above narrative with the reader because I learned a great deal in how the student teacher, Ella, responded to the moment and how she thought about it as a problem. Ella did not notice—on her own—the opening into zero, into negative numbers, or into children's developing understanding of what it means to subtract. This is not due to any fault of hers; rather, she consumed school mathematics in the Canadian school system in the 1980s when the thrust of mathematics in schools was the "rote recitation of facts, memorization of algorithms, and solving of routinized word problems" (Friesen, 2008, p. 139), much as I experienced it as a student. The exploration of mathematics in world political history was not one of the learning objectives in the 'back to basics' approach of the late 1960s, 1970s, and 1980s. In talking about the opening into zero, Ella wondered how to set up the activity in the future so that she could avoid the possibility of such a mathematical problem—a dangerous idea—from ever 'happening' again with children of this age. The flattening out of the classroom landscape in an attempt to make everything smooth and easy and predictable is one way to deaden a vibrant living discipline, but that was Ella's immediate response. It was a way for her to fix the problem. She was looking for what Caputo (1987) calls "the easy way out" (p. 3) by what Dunne (1993) describes as the lure of technique. I understand why Ella would want to make things easier. I do not blame her response as though she holds some deficit as a student. She responded in a way that I am sure that I too have responded in my own teaching, as the very nature of schools as they are currently structured in Western culture fosters fragmentation. My job as Ella's university instructor was to try to help her make her way to Dunne's (1993) rough ground and to see the problem differently.

I needed to point to the moment and help unearth for her and her fellow students the tensions that existed around such an opening. I remind readers that it was while working with students like Ella that I was first drawn to my topic of wonder for this book. Understanding begins "when something addresses us" (Gadamer, 1989, p. 299). The university students' responses to mathematics and their past experiences with mathematics in schools addressed me and would not let go. I was led on to think about the kind of mathematics that exists in schools and about the difficulties that such a view of mathematics provides in coming to understand the teaching of mathematics to children.

During my work with the education students we were able to spend time with one another, with children, and practicing teachers in schools as the narrative above briefly shows, and together we tried to make sense of the difficult and messy work of teaching children mathematics. Such experiences—the boys' announcement that $100 - 101 = 0$—addressed me. Being able to stand beside student teachers as we witnessed children working with mathematics and other disciplines meant we could unpack living cases together back in our own university classroom space. We could try to make sense of what we were seeing and hearing. Working in a teacher preparation program that was intentionally structured in this way meant that university students and instructors could have ongoing conversations about theory and practice. We could try to make sense of what we were experiencing in schools. Children's questions and comments made their way into our university classroom. We discussed what it meant to have practical wisdom or phronesis, a term first coined by Aristotle. Gadamer (1989) writes, "Practical knowledge, phronesis, is another kind of knowledge. Primarily, this means that it is directed towards the concrete situation" (p. 21). Phronesis is concerned with the actions that we take in particular situations with an understanding of how the particular speaks to general notions. Towers (2007) describes it as "a particular kind of knowledge—one oriented to action, and specifically ethical action, action oriented to the good" (¶ 3). Practical wisdom in teaching is a kind of knowledge that contrasts with techne, a knowledge that is technical in nature, and to which many of my education students initially were drawn. Phronesis is knowing what to say and what not to say. It is knowing when to speak and when to be silent. It is about recognizing when action is needed, and when no action is the best action. In some ways this is connected to van Manen's (1990, 1991) ideas regarding pedagogic tact, another important teaching element that we strived to better understand and work toward. Spending time with primary/elementary education students in a teacher education program that required my presence in schools and university classrooms was challenging for me as an instructor, but now as I look back to that time spent with student teachers, walking beside—dwelling—with student teachers in schools and in university classrooms was essential in helping both the university students and me come to a better understanding. We could journey together, and it provided opportunities for the education students to travel the mathematical landscape with someone slightly more experienced in that world with children, someone who might notice an opening to wonder and who could help them make some sense of what they were witnessing.

After working for three years with pre-service teachers focusing on elementary education preparation in western Canada, I accepted a new position at a post-secondary institution in eastern Canada, where part of my academic task would be to continue to work with pre-service primary/elementary generalists as they grapple with what it means to begin to understand the teaching and learning of mathematics for primary- and elementary-aged children. I was excited to begin this new academic position.

Looking back, I believed that my challenge was going to be easier in the new position since the students in the mathematics methods classes I would be teaching had to complete a minimum of two university mathematics courses for acceptance into their education program. Because of this programmatic fact, I had hope that my new students would know their way around the mathematical landscape a bit more than my former students who did not necessarily have strong university mathematical backgrounds. I knew I would still have to help them challenge their own assumptions about children and schooling. I just thought that perhaps I would not need to spend as much time helping them to explore the mathematical landscape. I was wrong. In chapter 3 I first mentioned alethia, the Greek word for 'truth.' In hermeneutics, it can mean un-concealment, uncovering, and opening up. Alethia requires others to be reminders of our forgetfulness. The pre-service teachers reminded me that I had forgotten about how the impact of their years of being in schools with a particular kind of mathematics had formed who they had become, as the following narrative will show.

Why Didn't Anyone Tell Me?

8:59 a.m. I got to the classroom just in time to find a decent piece of chalk, a small treasure in many university classrooms. On the table at the front of the room sat piles of carefully placed colored papers. A small pile of children's scissors stuck out of an old tin can with the soup label torn off. I went out to the hallway. Students were still weaving their way through the corridor, scurrying to make classes in time for their 9 a.m. starts. Already our classroom was crowded but I could see more students making their way, many out of breath after climbing flights of stairs. I went back into the classroom. Chairs scraped on the well-worn floor, and the sounds of zippers and Velcro mixed with "Good morning" and "How are you?" and the rustling of papers as notes from other courses were passed between students. I looked at the now forty pairs of eyes looking back at me. It was time to begin.

9:25 a.m. The increasing chatter in the room was a good indication that groups were nearing the end of the construction of their kits. Some of the more conscientious students were picking up scraps of paper and returning borrowed scissors. I had asked the students to pay attention to what they noticed regarding patterns and relationships between the various fraction pieces they were constructing. After a brief conversation regarding the various strategies that groups used, it was time to present the class with a problem.

Grabbing the chalk, I turned and began to write on the chalkboard. Instantly I could hear the scramble for pens and paper. [I always find it interesting how what gets *said* in class is not nearly as important as what gets pinned down in words on a board.] With the dusty chalk, I wrote the following problem for all to see: $1\frac{3}{4}/\frac{1}{2} = x$.

I stood back from the problem, making sure that I had written what I had intended to write and that it was legible. Turning back around, there were forty silent students staring at me. It was these students' first class for the week and I could tell some were still not quite alert, despite spending the past twenty minutes or so cutting paper into strips of various sizes. Almost empty Tim Horton's cups peppered the room, but the scent of coffee still wafted through the air.

So, one and three-quarters divided by one-half is what?

Immediately pens went to paper, scribbled on the sides of old notes, on top of the scraps created in the cutting out of the fraction pieces, on the sides of fraction pieces. I interrupted this.

Drop your pencils for a moment. How many of you have a sense of what x might be?... Do you have an idea of 'about how many' x represents?

No response.

Okay. Continue to solve the problem in whatever way you wish.

They carried on scribbling for a few more seconds before looking up once more.

Who would like to share their answer for this problem?

Mike, one of three men in the class, raised his hand.

I do, Miss. It's 3 and 1 over 2.

So what did you do?

Our group just did it. Got 3½.

Can you explain the process of coming to 3½?

Well, I 'timesed' 4 by 1 and added 3 and that got me 7 over 4 and then flipped the ½ to be 2 over 1 and 'timesed' the 7 over 4 by 2 to get 14 over 4 and then divided the 4 into 14 and got 3 and 2 over 4 and then reduced the 2 over 4 to ½.

Would anyone like to challenge Mike on his answer?

Silence.

Did anyone get an answer that differed from 3½?

Continued silence. Squirming in seats. Eye contact actively avoided.

So how many of you here in the room had a good idea of what roughly the answer would be when looking at the problem?

Silence.

What I mean is, that when you looked at it you knew it would be 'around 3 and something.'

Silence. Big sighs in the room. I looked to the back, and noticed a frustrated Monica, who saw her cue to speak:

I had no clue. I hated doing fractions. I just memorized the rule. I have no idea how I got 3½. I just did it. Flipped and multiplied. That's all there is to it. Memorize the rule and you'll get it right. I got the same answer as Mike.

Okay. Well, what is the problem really asking of us?

Grabbing the chalk again, I wrote another problem underneath my first one. 9/3 = x.

If I wrote the two rational numbers as whole numbers, like these, 9/3 = x what does the problem ask of us?

It's asking how many times 3 goes into 9, Miss.

How many 3's are in 9?

As I spoke I used my hand to point out the movement from the 3 to the 9.

So let's look at our initial problem... 1¾/½ = x. How many ½'s are in 1¾?

I picked up a circular fraction kit model that I had placed on the table next to the board and held up a circle made up of four quarter pieces and then three more quarters of a circle to represent 1¾. Then I took a ½ piece of a circle of the same size and laid it on top of the first representation. As I did this, the students counted in unison:

1, 2, 3, and a half.

Throughout the room whispers of "Oh wow! I see that. That's cool!" could be heard as some students explained again how it worked to a neighboring student at their table. But Monica was still not convinced.

But why is that a half? Isn't it a quarter?

Great question, Monica.

This was a great question. I paused and looked around the room.

How might we respond to what Monica has asked?

Jill, seated up near the front of the room, turned to face Monica.

The thing I did was to try to remember the question like if it was 'how many 3's are in 9.' I thought, How many semicircles are in that whole circle with the three-quarters of a circle? So the question is really asking how many ½'s can fit. Three halves and then half of a half circle. So 3½ halves.

Suddenly Monica opened her mouth wide. Her eyes became big.

OHHHHHHH!!! That is so cool!!!!!

This was a big moment. Monica had been fighting with her self-perceived mathematical inferiority since the beginning of the semester. For her to consider something mathematical as being 'cool'—such as the division of fractions—was a giant step.

So, how would you do this problem, using the fraction kits you just created?

Immediately they began laying the pieces out and working on the problem. Then they began to ask for more problems to try. Before we knew it, our class time was up for the day. Already the next class was hovering at the door, waiting to get settled for their class. But some of the students remained in their seats. They appeared almost emotional. It caught my attention enough for me to come and sit down next to them and ask whether everything was okay. They displayed intrigue at understanding better the division of fractions, but more than intrigue was the anger/frustration/embarrassment that it took this long in their education to figure out what they considered to be elementary mathematics. Jody sat there, slumped in his chair. The second class began to spill in and fill vacant chairs. With frustration he looked at me and said, *"So, was I supposed to learn this stuff on my own? Why am I only understanding this now, for the first time?"*

Jody's Why? question is not an uncommon one for me to hear in a mathematics education course as part of a primary/elementary teacher education program. Many of my students had been schooled in mathematics through a K–12 school system in which the normal emphasis was on rote learning and following procedures. The pedagogy most of them experienced was based on mathematics textbooks. It was a system with a fixed, fragmented, absolutist view of mathematics that was far removed from the living world. Students' statements such as Jody's are rife with implications of not being good enough and not knowing enough. Britzman (1991) states that such comments reveal fears of many prospective teachers who find themselves believing the myth of the teacher as the expert:

> One of the most commonly expressed fears of prospective teachers is that they will never know enough to teach. Two fears are collapsed into one: knowing how to teach and knowing everything there is to know about the material. (p. 227)

As the one whose job it is to help them learn about teaching mathematics to children, I often find it challenging to assuage their fears and insecurities. Having had some experience in working with pre-service teachers for the past ten years, I now realize that it is important to articulate clearly to them that there is nothing wrong with them because they are only now beginning to understand division of fractions. I try to help them see that we are products of a system in which a particular kind of mathematical knowing was fostered. As learners, they were to come to answers immediately and answers were either right or wrong, and discovered by following the rules using a single method. I try to help them uncover that their mathematical knowing and being have been fostered by a system—and a mathematics—that is culturally, politically, economically, pedagogically, and socially value laden.

Mathematics Autobiographies: Living with the Cover Version

From my initial work with pre-service teachers that led me to my topic of coming to understand the teaching of mathematics to children, I have come to the awareness that how students have dwelled with mathematics has formed who they have become, as learners and prospective teachers of mathematics. Part of my task, then, is to help them to challenge the assumptions they hold about what mathematics is for them, who they perceive themselves to be as learners of mathematics, and who they might become as teachers of mathematics, as well as teachers of the other disciplines for which they will be responsible.

To help my students begin to reflect on themselves as mathematics learners and future teachers, I ask them to write a mathematics autobiography at the beginning of their mathematics education course. Such an activity has many pedagogic purposes. It allows me as their professor to get to know my students. I learn about their school experiences and how such experiences have shaped their feelings about the subject. I also learn what students think about themselves as mathematics learners. Asking the students to reflect on how their lives have intersected with mathematics forces them to begin to unpack why they feel the way they do about mathematics and themselves as learners and future teachers. The students share their autobiographies with one another and the whole class. These conversations open up many avenues to be explored. Issues of gender, social status, and power emerge. Students find it remarkable that so many of them have common mathematics stories. They tell of remembering the colors and the picture on the cover of their grade-four mathematics textbook but cannot describe a moment of mathematical wonder from kindergarten to grade six. As prospective primary/elementary generalists, many students really do not like mathematics at all, despite having taken mathematics all through their schooling and they are the ones who have come out of the mathematical 'sorting machine' as the privileged ones. My students in eastern Canada have passed a required mathematics placement test for acceptance to university and have even gone on to complete two required university mathematics courses as a prerequisite for their education program. But many of them have been living with a certain version of knowing mathematics throughout roughly sixteen years of schooling. They have managed successfully to complete their school requirements but many still have a limited understanding of mathematics by their own claims and as most of their autobiographies can attest. They can follow procedures, such as flip and multiply when dividing fractions, but tell me they do so with little conceptual understanding. I refer to this version of knowing as the cover version of mathematics. As pre-service teachers these students will one day be responsible for opening up the world of mathematics to children and showing them how to be in the world with mathematics. It seems that their fears and insecurities about teaching and learning mathematics well up and spill over whenever our work together exposes them to new understandings of mathematics.

Boaler (2008) states that "mathematics is widely hated among adults because of their school experiences and most adults avoid mathematics at any cost" (p. 4). How has it come to be that a discipline deemed so important and privileged in society has produced so many people who now loathe the subject and avoid it as much as possible? How has it claimed such a widespread negative image? Ernest (2008) writes that the image of mathematics in the Western culture is one that "is difficult, cold, abstract, theoretical, ultrarational, but important and largely masculine" (p. 5). Could the very positioning of this image of mathematics and science at the top of the hierarchal structure of school curricula worldwide be the very thing that has shaped how so many students feel about it? I remind readers that I am referring to people like many of the students in my primary/elementary education classes who were 'successful' in school mathematics. It is important to note that currently the majority of my education students are women. It would be irresponsible of me not to help them peel back some of the issues of gender and let them in on how the kind of mathematics to which they have been exposed has positioned many of them as learners. The self-reflection that the activity of the autobiography assignment draws out begins some of these conversations. However, the short amount of time together in the classroom does not allow me to do this well with them.

I return to my initial belief that taking more university mathematics would help my students recognize the mathematical terrain. For some education students, yes, this is the case. But for many of them, this requirement might be problematic. Noddings (1997) questions whether the coercion of children to study mathematics in school has made it poisonous somehow. She writes that the forced taking of mathematics is not unlike going to the dentist.

> You do not always enjoy going but at least you can count on getting your teeth fixed. Rather, 'Taking' mathematics does not guarantee learning mathematics. When we force students to take mathematics, we can only ensure that they will have certain credentials to present; we cannot ensure that they have learned the mathematics required for the next level. They may or may not learn skills, concepts, and ways of thinking. As John Dewey argued so persuasively, people are far more likely to learn when they are thoroughly occupied and engaged in what they are doing and when they have participated in the construction of their own learning objectives. (p. 328)

In considering this, I have come to realize that for many of my primary/elementary education students, the greater number of experiences with mathematics that are required of them do not necessarily help them become open to the rich mathematical landscape that children deserve to explore. Rather, for many such education students more experiences with a mathematics that is to be endured, a mathematics severed from their world, might strengthen their distain and push them further from seeing possibilities of wonder with children on the mathematical terrain.

I do not need to look far to find claims about how the nature of mathematics in schools shapes how we feel about mathematics and ourselves as learners. I turn to Hersh (1997), whose work as a philosopher and mathematician has contributed to a reconceptualized mathematics. Hersh states that mathematics is misrepresented in schools, and that is why many people—like most of my education students—do not like mathematics. Focusing on their work with pre-service teachers in central Canada, Gadanidis and Namukasa (2007) state that school-like mathematics is "the kind of mathematics that turned them [pre-service teachers] off the subject in the first place" (p. 14). Boaler (2008) writes: "The math that millions of Americans experience in school is an impoverished version of the subject and bears little resemblance to the mathematics of life or work or even the mathematics in which mathematicians engage" (pp. 15–16). Friesen (2008) makes much the same claim, stating, "There is no other discipline in which the gap between the curriculum presented to students and the body of knowledge that constitutes the discipline is so great" (p. 137).

I wish to focus on Boaler's (2002) contributions at this point, as her work has provided significant findings to the field of mathematics education research. Boaler suggests that a great deal of what has shaped who we are, as mathematics learners, has been determined by the teaching and learning experiences we have had in schools. Schoenfeld (2002) explains Boaler's groundbreaking study this way:

> She interviewed the students about a wide range of topics, obtaining information that would allow her to construct portraits of the students' knowledge, beliefs, and mathematical identities. These qualitative, anthropological observations were mixed with a wide range of quantitative studies—extensive analyses of student performance on standardized exams and on specially constructed tasks that focused on aspects of problem solving left untapped by the standardized exams.... Boaler's multiple sources of evidence and triangulation on the data provide a robust and well-documented body of findings. (p. x)

In this award-winning research study, Boaler (2002) examined two vastly different approaches to the teaching and learning of mathematics. Working with two schools in England with differing philosophies but almost identical demographics for both the teachers and students, she explored multiple aspects of the experiences of students and teachers. The impact of how the students experienced mathematics dramatically influenced their beliefs about who they were as mathematics learners, the kinds of mathematical knowledge they held, and their understandings of mathematics in their world. One school, Amber Hill, had a pedagogy that was textbook driven in its teaching approach. It could be the archetype for the 'traditional school.' Its teachers, with strong mathematics backgrounds, focused on procedure and following rules, or mathematics along a rutted path (Bransford et al., 2003). This school made use of ability grouping in all its classes.

The other school, Phoenix Park, was project based with a generative curriculum in which students worked on engaging, ill-defined, open problems—mathematics-as-landscape; it would fit the archetype of a 'reform-based school.' Students were invited to explore the mathematical terrain with their teachers, who had mathematical backgrounds similar to the teachers at Amber Hill. Classes at Phoenix Park were grouped by mixed ability.

Boaler's findings have significance. Overall, the students who had the opportunity to explore the mathematical terrain through the project-based approach did significantly better in a variety of assessments, including the General Certificate of Secondary Education (GCSE) examination for mathematics. Students were able to transfer their understanding to new situations and solve problems that were unfamiliar to them. The data revealed the students had confidence in themselves as mathematics learners. Boaler (2002) writes:

> The students were clear in interviews about the source of their mathematical confidence. They related this to two features of their approach: The fact that they had been forced to become autonomous learners, and that they had always been encouraged to think for themselves. (p. 127)

The research data showed that students at Phoenix Park, both male and female, appeared to enjoy mathematics.

In contrast, the students at Amber Hill did not fare nearly as well on either their formal examinations or applied assessments. The students struggled to transfer their mathematics knowledge to new situations. Their attitude toward mathematics was that it was a subject to be endured and it did not make sense, and they did not enjoy it. Many students lacked confidence in understanding mathematics and the consequences were the direst for female learners at the traditional school.

Stigler and Hiebert (1999) analyzed videotaped mathematics lessons from three countries—Germany, the United States, and Japan—and they write about the type of teaching approach that is characterized by the teachers at Amber Hill, in which learners "spend most of their time acquiring isolated skills through repeated practice" (p. 11). Boaler (2002) states that the type of approach used at Amber Hill is far from unique, as Stigler and Hiebert's study also indicates. Boaler (2002) describes this approach as "a deeply cultural phenomenon pervasive in the Western world" (p. 49).

The type of mathematical knowing that many students from Amber Hill were coming away with is a form of the cover version of knowing similar to many of my university students who wish to become primary/elementary teachers. Hard-working and dedicated teachers covered the mathematics curriculum. Students passed from grade to grade, successfully completing unit tests and final exams, and yet mathematics did not make much sense to many of them. This type of knowing has led many education students who strive to become primary/elementary generalists to see mathematics as absolute, rigid, and disconnected. Indeed, such experiences for mathematics learners in schools have contributed greatly to shaping who they are by the time they land in my mathematics methods classes.

Hermeneutics leaves me with the ethical task of deciding how to properly proceed. When pre-service students tell me they dislike mathematics but want their future students to like the discipline, I have a responsibility to them. I need to help them become critical about the Western, Eurocentric mathematics they experienced and to be critical about the type of mathematics they will recover for their future students. I have to help them see that they have been consumers of a type of mathematics that is filled with inequality. Christensen, Stentoft, and Valero (2008) write about ethnomathematics as challenging the reign of "Western white mathematics," and they state:

> Through the historic development of the West—which has a well-documented impact on the transformation of people in other parts of the world—mathematics has imposed the rationality of the dominant power over other ways of thinking and expression in non-Western, indigenous, colonized cultures. (p. 143)

Prospective teachers need to be aware of other ways of knowing mathematics and that mathematics can exist in the world as a form of oppression. Keitel and Vithal (2008) explain further:

> The question of who will be the producers and who will be the different kinds of users of mathematics becomes important because mathematical power, through a science and technology driven knowledge and skills market economy then participates in a particular distribution of other kinds of (political or economic) power. Large numbers

of people also get left completely outside mathematics and its many products made available through science and technology—the poor, the homeless, those caught in conflicts and wars, those who do not get access to education and so on. (p. 175)

Social justice, feminist, and critical mathematics education pedagogies can help orient students to another way of being with mathematics and I realize that I need to spend more time investigating these pedagogies carefully. In this book I have paid attention to a pedagogy that is inquiry based; such a pedagogy can allow for social justice, feminist, and critical perspectives. Certain ways of teaching and learning such as inquiry-based teaching and learning are more ontological in their dispositions, as I tried to show in chapter 4 in an inquiry-based classroom. Ontologically speaking, you become someone because of what you know. Understanding turns us into people who conduct ourselves in different ways. For future teachers to begin to see the difficulty and complexity of teaching mathematics, it is important for teacher education programs to attend to the rough ground of teaching as "indefinite, indeterminate, complex, contextual and always mutable" (Phelan, 2003, ¶ 4). This is difficult to do in a teacher education program that is course driven and void of opportunities for learners to grapple with theory and practice interchangeably. Such a teacher education program perpetuates fragmentation. Regardless, as teacher educator I need to help future teachers in whatever ways possible to attend to the rough and wonderful terrain of mathematics. I need to provide students with new images of what is possible in schools to counter the dominant image of teaching and learning through textbooks and standardized test–driven political agendas. I also need to remind students that they are just beginning their journey as students of teaching and learning. Remembering that their journey will never end is important for their growth as students of teaching and learning. Reminding them to return to the ontological question Why? around all they think about and do as they learn about the nature of mathematics, schools, and curriculum is paramount. I too need to turn to face Why? 'So, why is any of this important?' The concluding chapter of this book will attempt to sit with this question.

· 9 ·

RESPONSIBILITY TO RECOVER

You Can't Get There from Here: Returning Home

When I speak with visitors to the place that is now my home, they will sometimes feign gentle frustration at the good-natured response they might get from a local citizen when asking for directions to a place of interest.

"You want to go where? Oh. [*Pause.*] Well, [*pause*] you can't get there from here." Almost immediately after hearing this, the puzzled tourist is given detailed directions to get to the place in question, but the directions necessitate that the tourists go somewhere else first, before getting to their destination. They must journey elsewhere, and possibly to a place they had not desired to go, before being able to arrive at their target.

The phrase 'You can't get there from here' reminds me also of the writing of a hermeneutic inquiry. During the writing of one of my earlier chapters, I had at one point naïvely considered jumping ahead to the end, to say what I thought I was figuring out as a way of guiding my work to that destination. Perhaps this temporary way of thinking was my solution to having a response when people unfamiliar with hermeneutic research would ask about my research findings. I did not jump to the end, however, because I was unable to do so. 'I could not get there from where I was.' I realize that to do so would not have been true to hermeneutic writing. I had to 'write myself' to the conclusion through the disruptions that one necessarily needs to bump into and acknowledge.

And so I am someone different from the person I was when I began this writing. I am ontologically changed by this work as this journey has also been a journey of a deepening self-understanding. What is it that I can conclude about my journey over the past hundred pages, and more important, why should the reader care? What does this writing mean for children, pre-service teachers, practicing teachers, teacher educators, or anyone who cares about children, schools, and mathematics? This chapter will attempt to respond to these questions. Also, I will provide a final look back to my grade-one and-two students who gave me so much as a way to pay tribute to them, my greatest teachers.

We Tell Our Stories to Find Out What They Mean

Gadamer's hermeneutics calls to me and his ontology of understanding advances my work because of what I recognize as a hermeneutics of possibility, of hope, of generosity, and of responsibility. Engaging in conversation, looking for what might be true and for what might be possible is indeed a generous offering. Gadamer (1989) describes the task of hermeneutics in this way: "What man [sic] needs is not just the persistent posing of ultimate questions, but the sense of what is feasible, what is possible, what is correct, here and now" (p. xxxviii). It has never been my intention to fix the problem with mathematics or to come to firm solutions for the best way to teach children. The purpose of this hermeneutic journey has been to allow me to examine my lived experience to make meaning of those experiences. While keeping the original difficulty alive (Caputo, 1987), I am working toward understanding the teaching of mathematics to children. David Jardine said a few years ago in his guest lecture at a research institute on hermeneutics, "We tell our stories to find out what they mean" (Inaugural Hermeneutics Institute, June, 2009). This hermeneutic inquiry has not just been a telling of stories, but more important, I believe these narratives have been telling of something—telling of possibilities with children, of life in classrooms, of the difficulty of teaching, of who we might become as teachers, of the nature of mathematics, of what it means to be a primary/elementary pre-service teacher, and of complexities surrounding mathematics education.

What it is that I have come to understand more fully, and what it is that I hope the reader has come to understand is this: Children deserve the chance to dwell with mathematics and be exposed to the wonder and beauty of the living discipline. They need to get the chance to play with mathematics authentically, using their whole bodies in lively, rich mathematical engagements. We have a responsibility to recover mathematics for children. We have a responsibility to return it to the world from which it has been severed. Children teach us so much about the world and ourselves; they do remind us of the simple joys of discovery and exploration; they do challenge us not to be so cynical.

I know that the type of mathematics to which I refer has its challenges. There are no Mad Minute tests for speed and accuracy. As a mathematics education scholar I know from research how harmful this is for learners. There are no answers at the back of the book. There may be no answers for some things at all, just yet. Children deserve to know that being human in a world with mathematics does not mean there are answers at the back of the book. It does not mean that problems come packaged, neat and tidy. Being human does not mean that things are isolated and severed from their histories and the world. Gadamer (1981, 1989) writes about human finitude. The one thing we can be sure about is that we will not live forever. We are in the process of becoming, just as mathematics is in a process of becoming, and have yet to become all that we are. Indeed, letting children in on what it means to be human invites them into an awareness of their own finitude. Children will gain self-knowledge and come to a better understanding of what it means to be human through their play with mathematics as an authentic, living, human enterprise. It will make their lives more difficult, more authentic, and more human. They will lead better lives in classrooms that take up mathematics in this way.

As teachers we need to attend to children, in all their imaginative, playful innocence, and we need to listen to what they ask of us as teachers. We need to have faith in children and in ourselves as we negotiate the difficult and delicate work of living well with children in schools. Sometimes we forget how powerful the messages are that we pass on to our students. Yeager et al. (2013) did a study that showed that achievement increased—particularly in minority students—when students perceived that their teacher really believed in them. Mueller and Dweck (1998) conducted a study that looked at the immediate impact that teachers' words have on students. In their study, half of the students were praised for being smart as they completed a given task. The other half were praised for working hard. When given another task to do, almost all (90%) of the children praised for being smart chose an easier task while the majority of children praised for working hard chose a more difficult task. Dweck (2006) continued to work on these ideas and wrote a *New York Times* best seller based on her years of research into learner mindsets. This research has huge implications for mathematics learning. According to Dweck (2006), there are two kinds of mindsets that learners can have—*a fixed mindset* or a *growth mindset*. With a fixed mindset the learner sees intelligence as static and math ability as something one either has or doesn't have. Students with a fixed mindset strive to appear smart and therefore avoid challenges, see effort as useless, feel threatened by the success of others, and give up easily. Interestingly, many high-achieving girls in mathematics suffer from a fixed mindset. Students with a growth mindset see intelligence as something that can be developed, leading to a desire to learn by embracing challenges, working hard, and persisting when faced with obstacles. Boaler (2013) states that in the United States more people have a fixed mindset toward mathematics than toward any other topic in their lives. She argues that this fixed mindset thinking about mathematics is one of the reasons for so much math trauma and failure. The good news is that people with a fixed mindset can actually learn to have a growth mindset and as teachers it really is our task not only to disrupt notions of what mathematics is, but also to focus on what it means to learn mathematics with a growth mindset. It is critical that as teachers and as parents we ourselves strive to have a growth mindset if we do not have one already. It is critical to praise effort with authenticity and not to foster the myth that you are either a 'math person' or you are not.

As mathematics teacher educators, we need to recover mathematics for our pre-service teachers and graduate students. They, too, need to realize that to be human means there are no answers in the back of the book and that they can learn mathematics through transitioning to a growth mindset. We must disrupt their desire for 'the easy way out' and help them challenge the myth of the teacher as expert (Britzman, 1991). It is not easy to convince other adults, our university students, that they need to embrace uncertainty, as Wheatley (2005) suggests, for it means that as educators we too need to embrace uncertainty as we find our way. Disrupting notions of what mathematics is takes work, and it takes courage. To help students re-conceptualize and re-imagine mathematics is not easy. I realize that teacher education programs differ across North America, but as academics, we have a responsibility to the discipline of teaching and to mathematics to restore the world to mathematics and mathematics to the world. We need to embrace and inquire more deeply into understanding other ways of knowing mathematics. We need to continue to challenge the political, social, cultural, economic grand narratives through our research and our teaching. None of this work is going to be easy, but all of this work is necessary.

A Final Look Back

It is only fitting that I end this book with a narrative piece about the grade-one and grade-two students who helped me learn to see the world as a place of possibility, a place of joy, a place of difficulty, and a place of hope. This book began with a conversation in a school hallway with a child who wished she had never been born, since she had to endure school as I had structured it. Jarring as her words were, I will be forever grateful for that conversation.

Personal Journal Entry, Fall 2006

Preparing to leave what had become home for me in the foothills of the Rockies was difficult, but the challenge of a new academic position located on the edge of the North Atlantic was beckoning with possibilities. The move across the country presented the opportunity to clear out closets, recycle un-necessary items, and purge articles rendered no longer useful. It was early summer.

As a strange twist of fate, it was the summer solstice and I found myself at a garbage site near my former home. The summer solstice, June 21, was the day that my grade-one students and I celebrated our work publicly with family, friends, and the local community. It was the day upon which great feasts and much celebration took place in medieval times, and as investigators into the life of that particular time period it was only fitting for our class to celebrate our work. Many years had since passed, and here I was, perched above a giant bin filled with broken furniture and other discarded items. With the help of a friend, I struggled to haul the round table from the back of a red pickup truck.

For years I had held on to the round table. It made its way from one classroom to another classroom, to a closet in the school, and finally to my own garage. It remained an artifact for me—an artifact of time spent, for ideas shared, for memories of joys, struggles, and accomplishments achieved by the imaginative children who sat around it with me daily for two years. I briefly contemplated shipping the large, homemade, legless table to the other side of the country with my other belongings. But it was no longer practical to haul something so unwieldy a few thousand miles simply for the memories. I needed to let it go.

With gratitude I cast my eyes over the brightly painted signs of the zodiac one last time and thought about those children who had taught me so much. What they had taught me was the real gift that I needed to hold on to, not the plywood and paint. Do they know what they have given me? Some day I need to let them know.

Letter Unsent: A Tribute

Dear Students of 2S,

As I write this letter, you are beginning your final year of high school. Some of you may not even remember my name or remember much of our work together so many years ago, but I shall remember each of you always, for you have impacted my life in ways I could never have imagined. I have done many things as a teacher since we last spent time together in conversation around our round table. I now work at a university where my job is to teach and to do research in a faculty of education. I have moved back east, not to the land of Anne, but nearby, and I have met many people in my work—students, teachers, parents, administrators, and other professors. Often I speak about you fondly as exemplars of what children are capable of doing in school and becoming as learners; you surface in my teaching and my writing. Does this surprise you?

While I was responsible for teaching you, you might not realize that you taught me, too. I have been searching for a way to thank you, and to express my gratitude. This letter is one attempt to do so.

You taught me several lessons during our two years together. Here are just some of the things you did for me: You taught me that children have something to say about how school can be imagined. I learned to listen to your questioning of the world because you never stopped asking good questions. You taught me that it is important for teachers to be playful and curious when working with children. I tried to suspend disbelief, even for a little while. For some of you I wonder whether you still think I could have real phone conversations on bananas from your lunch. For all the interesting work we did together on medieval life, Monty Python, Beowulf, the post office, the Odyssey, Greek mythology, the exploration of Mars, our snake, and robotics, I wonder whether all that you remember is that I took you on the roof and talked on bananas. And if that is all that you remember of the work we did, it does not matter, for you became children who believed in themselves. You became children who could confidently speak about your world and your work, and you made other people care, too. You made people, including me, think differently about mathematics and what you were capable of doing in grades one and two. You made other people, including me, think differently about technology, about teaching, and about the power of conversations to come to new understandings. You showed me how even the very young can solve very big problems. You brought tremendous joy to me each day we spent together and helped me not to take life so seriously all the time. The work we did was not always easy for me, but I gained such energy from you and your ability to be imaginative, to be playful, and to believe.

I wish you much happiness, as you complete your final year of school and make decisions about your future education. I hope you still have the curiosity for the world around you. I hope you still believe that anything is possible. Some day, perhaps, you will come to understand how thankful I am for having spent time at the same round table as you.

With love and gratitude, Ms. S

NOTES

Foreword

1. Peter A. Rubba, Head of the Department of Curriculum and Instruction, Penn State University, 1994–2000.

Chapter 7

1. I received more students in grade two, so not all twenty-eight were in grade one with me.
2. Olympus Mons is the largest volcano in the solar system. When the students saw this volcano on the map of Mars, they insisted that we include it in our section of terrain. It is three times the size of Mount Everest.
3. *Elaphe* is Latin for 'cornsnake,' so the children chose that name for him. The children did all the work to bring about adopting and caring for Elaphe. My teaching colleague and friend, Cheryl, was critical in leading the rich curriculum that Elaphe's presence provided for all of us.
4. SMARTBoard is an interactive whiteboard and is a product of SMART Technologies.
5. Mad Minutes are timed computation exercises wherein children are given 60 seconds to do as many computations as possible.
6. The Gathering Area is a special space in the school used for meetings and special events. Located in the center of the school, it holds the same function as a gathering area in a traditional aboriginal community.
7. Red Rover Red Rover was an online project from the JPL in California. We received funding from the school to take part in this worldwide school project. Most of the other participants were from older grades, but it seemed a perfect fit for our work. We ran into problems, however, with Internet permissions. Most of the time our school's firewall blocked access to others to be able to operate our robot on our terrain.

BIBLIOGRAPHY

Abram, D. (1996). *The spell of the sensuous: Perception and language in a more-than-human world.* New York, NY: Pantheon.

Alberta Learning. (1996). *Alberta program of studies for k-9 mathematics.* Edmonton, AB: Alberta Learning.

Applebaum, P. (2003). Critical thinking and critical mathematics education. Retrieved January 14, 2014, from http://gargoyle.arcadia.edu/appelbaum/encyc.htm

Boaler, J. (2002). *Experiencing school mathematics: Traditional and reform approaches to teaching and their impact on student learning.* Mahwah, NJ: Erlbaum.

Boaler, J. (2008). *What's math got to do with it? Helping children learn to love their least favorite subject and why it's important for America.* New York, NY: Viking Penguin.

Boaler, J. (2013). Unlocking children's math potential: Five research results to transform math learning. Retrieved January 29, 2014, from http://youcubed.org/pdfs/teacher%20 article%20youcubed.pdf

Bransford, J., Brown, A., Anderson, J., Gelman, R., Glaser, R., Greenough, W. et al. (2000). *How people learn: Brain, mind, experience, and school* (Expanded ed.). Washington, DC: National Academy Press.

Britzman, D. (1991). *Practice makes practice: A critical study of learning to teach.* Albany, NY: SUNY Press.

Brown, S., & Walter, M. (2004). *The art of problem-posing* (3rd ed.). Mahwah, NJ: Erlbaum.

Caputo, J. (1987). *Radical hermeneutics.* Bloomington: Indiana University Press.

Christensen, O. R., Stentoft, P., & Valero, D. (2008). A landscape of power distribution. In E. de Freitas & K. Nolan (Eds.), *Opening the research text: Critical insights and in(ter)ventions into mathematics education* (pp. 147–154). New York, NY: Springer.

Clifford, P., & Friesen, S. (2003). A curious plan: Managing on the twelfth. In D. Jardine, P. Clifford, & S. Friesen (Eds.), *Back to the basics of teaching and learning: Thinking the world together* (pp. 15–36). Mahwah, NJ: Erlbaum.

Clifford, P., & Friesen, S. (2008). A curious plan: Managing on the twelfth. In D. Jardine, P. Clifford, & S. Friesen (Eds.), *Back to the basics of teaching and learning: Thinking the world together* (2nd ed., pp. 11–30). New York, NY: Routledge.

Clifford, P., Friesen, S., & Jardine, D. (2001). The ontology of hope: Lessons from a child. *Journal of Curriculum Theorizing, 17*(4), 59–66. Retrieved August 20, 2009, from http://www.udel.edu/aeracc/papers/The%20Ontology%20of%20Hope%202001.html

Davis, B. (1995). Why teach mathematics? Mathematics education and enactivist theory. *For the Learning of Mathematics, 15*(2), 2–8.

Davis, B. (1996). *Teaching mathematics: Toward a sound alternative.* New York, NY: Garland.

Davis, P., & Hersh, R. (1986). *Descartes' dream: The world according to mathematics.* Boston, MA: Houghton Mifflin.

Dewey, J. (1938/1998). *Experience and education: The 60th anniversary edition.* West Lafayette, IN: Kappa Delta Pi.

Dunne, J. (1993). *Back to the rough ground: "Phronesis" and "techne" in modern philosophy and in Aristotle.* Notre Dame, IN: University of Notre Dame Press.

Dweck, C. (2006). *Mindset: The new psychology of success.* New York, NY: Ballantine.

Enzensberger, H. (1997). *The number devil: A mathematical adventure.* New York, NY: Henry Holt.

Ernest, P. (1996). The nature of mathematics and teaching. *Philosophy of Mathematics Education Newsletter, 7.* Retrieved January 30, 2014, from http://www.ex.ac.uk/~PErnest/pome/pompart7.htm

Ernest, P. (2003). Conversation as a metaphor for mathematics and learning. *Philosophy of Mathematics Education Newsletter, 17.* Retrieved January 4, 2014, from http://people.exeter.ac.uk/PErnest/pome17/metaphor.htm

Ernest, P. (2008). Epistemology plus values equals classroom image of mathematics. *Philosophy of Mathematics Education Journal, 23.* Retrieved January 25, 2014, from http://people.exeter.ac.uk/PErnest/pome23/index.htm

Friesen, S. (2000). *Re-forming the mathematics in mathematics education* (Unpublished doctoral dissertation). University of Calgary, Alberta, Canada.

Friesen, S. (2008). Math: Teaching it better. In D. Jardine, P. Clifford, & S. Friesen (Eds.), *Back to the basics of teaching and learning: Thinking the world together* (2nd ed., pp. 135–142). New York, NY: Routledge.

Friesen, S., Clifford, P., & Jardine, D. (1998). Meditations on classroom community, memory and the intergenerational character of mathematical truth. *Journal of Curriculum Theorizing, 14*(3), 6–11.

Friesen, S., Clifford, P., & Jardine, D. (2003). Jenny's shapes. *Philosophy of Mathematics Education Newsletter, 17.* Retrieved January 15, 2014, from http://www.ex.ac.uk/~PErnest/pome17/pdf/jenny.pdf

Gadamer, H. (1981). *Reason in the age of science*. Cambridge, MA: MIT Press.

Gadamer, H. (1989). *Truth and method* (2nd ed.). New York, NY: Continuum.

Gadanidis, G., & Namukasa, I. (2007). Mathematics-for-teachers (and students). *Journal of Teaching and Learning*, 5(1), 13–22.

Gordon Calvert, L. (2001). *Mathematical conversations within the practice of mathematics*. New York, NY: Peter Lang.

Greene, M. (1995). *Releasing the imagination: Essays on education, arts, and social change*. San Francisco, CA: Jossey-Bass.

Heidegger, M. (1977). *Basic writings*. New York, NY: Harper & Row.

Hersh, R. (1997). *What is mathematics, really?* Oxford, England: Oxford University Press.

Jardine, D. (1994). The stubborn particulars of grace. In B. Horwood (Ed.), *Experience and the curriculum* (pp. 261–275). Dubuque, IA: Kendall/Hunt.

Jardine, D. (1998). *To dwell with a boundless heart*. New York, NY: Peter Lang.

Jardine, D. (2006). The fecundity of the individual case: Considerations of the pedagogic heart of interpretive work. In D. Jardine, S. Friesen, & P. Clifford (Eds.), *Curriculum in abundance* (pp. 151–168). Mahwah, NJ: Erlbaum.

Jardine, D. (2008). The profession needs new blood. In D. Jardine, P. Clifford, & S. Friesen (Eds.), *Back to the basics of teaching and learning: Thinking the world together* (2nd ed., pp. 195–210). New York, NY: Routledge.

Jardine, D., Bastock, M., George, J., & Martin, J. (2008). Cleaving with affection: On grain elevators and the cultivation of memory. In D. Jardine, P. Clifford, & S. Friesen (Eds.), *Back to the basics of teaching and learning: Thinking the world together* (2nd ed., pp. 31–58). New York, NY: Routledge.

Jardine, D., Clifford, P., & Friesen, S. (Eds.). (2003). *Back to the basics of teaching and learning: Thinking the world together*. Mahwah, NJ: Erlbaum.

Jardine, D., Clifford, P., & Friesen, S. (2008). Introduction: An interpretive read of "back to the basics." In D. Jardine, P. Clifford, & S. Friesen (Eds.), *Back to the basics of teaching and learning: Thinking the world together* (2nd ed., pp. 1–10). New York, NY: Routledge.

Jardine, D., & Friesen, S. (2008). A play on the wickedness of undone sums, including a brief mytho-phenomenology of "x" and some speculation on the effects of its peculiar absence in elementary education. In D. Jardine, P. Clifford, & S. Friesen (Eds.), *Back to the basics of teaching and learning: Thinking the world together* (2nd ed., pp. 131–135). New York, NY: Routledge.

Jardine, D., Friesen, S., & Clifford, P. (2003). Behind every jewel are three thousand sweating horses: Mediations on the ontology of mathematics and mathematics education. In E. Hsebe-Ludt & W. Hurren (Eds.), *Curriculum intertext: Place, language, pedagogy* (pp. 39–50). New York, NY: Peter Lang.

Keitel, C., & Vithal, R. (2008). Mathematical power as political power: The politics of mathematics education. In P. Clarkson & N. Presmeg (Eds.), *Critical issues in mathematics education: Major contributions of Alan Bishop* (pp. 167–188). New York, NY: Springer.

King, J. (1992). *The art of mathematics*. New York, NY: Plenum Press.

Kohn, A. (1999). *The schools our children deserve*. New York, NY: Houghton Mifflin.

Lakoff, G., & Núñez, R. (2000). *Where mathematics comes from: How the embodied mind brings mathematics into being*. New York, NY: Basic Books.

Mlodinow, L. (2001). *Euclid's window*. New York, NY: Simon & Schuster.

Mueller, C., & Dweck, C. (1998). Praise for intelligence can undermine children's motivation and performance. *Journal for Personality and Social Psychology, 75*(1), 33–52.

Noddings, N. (1997). Does everybody count? Reflections on reforms in school mathematics. In D. Flinders & S. Thornton (Eds.), *The curriculum studies reader* (pp. 324–336). New York, NY: Routledge.

Palmer, P. (1998). *Courage to teach*. San Francisco, CA: Jossey-Bass.

Palmer, R. (1980). The liminality of Hermes and the meaning of hermeneutics. Retrieved January 12, 2014, from http://www.mac.edu/faculty/richardpalmer/liminality.html

Palmer, R. (1999). The relevance of Gadamer's philosophical hermeneutics to thirty-six topics or fields of human activity. Retrieved January 22, 2014, from http://www.mac.edu/faculty/richardpalmer/relevance.html

Pappas, T. (1989). *The joy of mathematics: Discovering mathematics all around you*. San Carlos, CA: World Wide.

Paulos, J. (1991). *Beyond numeracy*. New York, NY: Vintage.

Phelan, A. (2003). Melancholia. *Educational Insights, 8*(2). Retrieved August 25, 2009, from http://www.ccfi.educ.ubc.ca/publication/insights/v08n02/contextualexplorations/provoke/melancholia.html

Postman, N., & Weingartner, C. (1969). *Teaching as a subversive activity*. New York, NY: Dell.

Prensky, M. (2001). Digital natives, digital immigrants. *On the Horizon, 9*(5). Retrieved January 15, 2014, from http://www.marcprensky.com/writing/Prensky%20-%20Digital%20Natives,%20Digital%20Immigrants%20-%20Part1.pdf

Sam, L. (2002). Public images of mathematics. *Philosophy of Mathematics Education Journal*. Retrieved January 15, 2014, from http://people.exeter.ac.uk/PErnest/pome15/lim_chap_sam.pdf

Scardamalia, M., & Bereiter, C. (2006). Knowledge building: Theory, pedagogy, and technology. In K. Sawyer (Ed.), *Cambridge handbook of the learning sciences* (pp. 97–118). New York, NY: Cambridge University Press. http://ikit.org/fulltext/2006_KBTheory.pdf

Schoenfeld, A. (2002). Foreword. In J. Boaler (Ed.), *Experiencing school mathematics: Traditional and reform approaches to teaching and their impact on student learning* (pp. ix–xiii). Mahwah, NJ: Erlbaum.

Seife, C. (2000). *Zero: The biography of a dangerous idea*. New York, NY: Penguin.

Sierpinska, A. (2004). Research in mathematics education through a keyhole: Task problematization. *For the Learning of Mathematics, 24*(2), 7–15.

Sinclair, N. (2001). The aesthetic is relevant. *For the Learning of Mathematics, 21*(1), 25–32.

Smith, D. (1991). Hermeneutic inquiry: The hermeneutic imagination and the pedagogic text. In E. Short (Ed.), *Forms of curriculum inquiry* (pp. 187–209). Albany, NY: State University of New York Press.

Smith, D. (1994). *Pedagon: Meditations on pedagogy and culture*. Bragg Creek, Alberta, Canada: Makyo.

Smith, D. (2006). *Trying to teach in a season of great untruth: Globalization, empire and the crises of pedagogy*. Rotterdam, The Netherlands: Sense.

Stewart, I. (2006). *Letters to a young mathematician*. New York, NY: Basic Books.

Stigler, J., & Hiebert, J. (1999). *The teaching gap*. New York, NY: Free Press.

Taylor, P. (1997, February 6). Needed: Smarts and small napkins. *The Globe and Mail*. Facts and Arguments.

Taylor, P., & Sinclair, N. (2000). Reinventing the teacher. *International Congress of Mathematics Education*. Retrieved August 5, 2005, from http://www.mast.queensu.ca/~peter/pdf/icme2000.pdf

Thom, J. (2003). What were they thinking? Exploring three children's mathematics. *Proceedings of the 2003 Complexity Science and Educational Research Conference*. Retrieved August 18, 2009, from http://www.complexityandeducation.ualberta.ca/conferences/2003/Documents/CSER_Thom.pdf

Towers, J. (2007). Using video in teacher education. *Canadian Journal of Learning and Technology, 33*(2). Retrieved January 22, 2014, from http://www.cjlt.ca/index.php/cjlt/article/view/7/7

Van Manen, M. (1990). *Researching lived experience*. Albany, NY: SUNY Press.

Van Manen, M. (1991). *The tact of teaching*. Albany, NY: SUNY Press.

Wheatley, M. (2005). *Finding our way: Leadership for an uncertain time*. San Francisco, CA: Berrett-Koehler.

Winning, A. (2002). Homesickness. Retrieved January 31, 2014, from http://www.phenomenologyonline.com/sources/textorium/winning-anne-homesickness/

Yeager, D., Purdie-Vaughns, V., Garcia, J., Apfel, N., Brzustoski, P., Master, A., et al. (2013). Breaking the cycle of mistrust: Wise interventions to provide critical feedback across the racial divide. *Journal of Experimental Psychology: General*.

INDEX

RETHINKING CHILDHOOD

GAILE S. CANNELLA, *General Editor*

Researchers in a range of fields have acknowledged that childhood is a construct emerging from modernist perspectives that have not always benefited those who are younger. The purpose of the Rethinking Childhood Series is to provide a critical location for scholarship that challenges the universalization of childhood and introduces new, reconceptualized, and critical spaces from which opportunities and possibilities are generated for children. Diverse histories and cultures are considered of major importance as well as issues of critical social justice.

We are particularly interested in manuscripts that provide insight into the contemporary neoliberal conditions experienced by those who are labeled "children" as well as authored and edited volumes that illustrate life and educational experiences that challenge present conditions. Rethinking childhood work related to critical education and care, childhood public policy, family and community voices, and critical social activism is encouraged.

For more information about this series or for submission of manuscripts, please contact:

Gaile S. Cannella
Gaile.Cannella@unt.edu

To order other books in this series, please contact our Customer Service Department at:

(800) 770-LANG (within the U.S.)
(212) 647-7706 (outside the U.S.)
(212) 647-7707 FAX

Or browse online by series at:
www.peterlang.com